Lecture Notes in Mathematics

Volume 2318

This series reports on new developments in all areas of mathematics and their applications - quickly, informally and at a high level. Mathematical texts analysing new developments in modelling and numerical simulation are welcome. The type of material considered for publication includes:

1. Research monographs
2. Lectures on a new field or presentations of a new angle in a classical field
3. Summer schools and intensive courses on topics of current research.

Texts which are out of print but still in demand may also be considered if they fall within these categories. The timeliness of a manuscript is sometimes more important than its form, which may be preliminary or tentative.

Titles from this series are indexed by Scopus, Web of Science, Mathematical Reviews, and zbMATH.

Jean Deteix • Thierno Diop • Michel Fortin

Numerical Methods for Mixed Finite Element Problems

Applications to Incompressible Materials and Contact Problems

 Springer

Jean Deteix
GIREF, Département de Mathématiques et
de Statistique
Université Laval
Québec, QC, Canada

Thierno Diop
GIREF, Département de Mathématiques et
de Statistique
Université Laval
Québec, QC, Canada

Michel Fortin
GIREF, Département de Mathématiques et
de Statistique
Université Laval
Québec, QC, Canada

This work was supported by Agence Nationale de la Recherche (ANR-18-IDEX-0001).

ISSN 0075-8434   ISSN 1617-9692   (electronic)
Lecture Notes in Mathematics
ISBN 978-3-031-12615-4   ISBN 978-3-031-12616-1   (eBook)
https://doi.org/10.1007/978-3-031-12616-1

Mathematics Subject Classification: 74S05, 65N22, 65F10, 65F08, 74B20, 74M15

This Springer imprint is published by the registered company Springer Nature Switzerland AG
The registered company address is: Gewerbestrasse 11, 6330 Cham, Switzerland

Contents

Chapter 1
Introduction

Mixed Finite Element Methods are often discarded because they lead to indefinite systems which are more difficult to solve than the nice positive definite problems of standard methods. Indeed, solving indefinite systems is a challenge : direct methods [4, 14, 78] might need renumbering (see [28]) of the equations and standard iterative methods [22, 81, 86] are likely to stagnate or to diverge as proved in [86]. As an example, consider the classical conjugate gradient method. Applied to a symmetric indefinite problem it will generate a diverging sequence. As the conjugate gradient method is (in exact arithmetic) a direct method, it will yield the exact solution if the problem is small enough to avoid loosing orthogonality. Applying a minimum residual method to the same problem will in most cases yield stagnation.

These two classical methods are the simplest in a list which grows constantly. This monograph does not intend to introduce new iteration methods. We shall rely mostly on existing packages, mostly Petsc from Argonne Laboratory [11].

Our concern is to solve algebraic systems associated to mixed discretisation. Several approaches (see, for example, [8, 15, 43]) exist in the literature to solve this type of problem but convergence is not always guaranteed. They are indefinite systems but also structured systems associated with matrices of the form,

$$\begin{pmatrix} A & B^t \\ B & 0 \end{pmatrix} \tag{1.1}$$

where A is often a positive definite matrix.

The key to obtain convergence is preconditioning. For general problem, a vast number of preconditioners are available [15]. Our goal is to develop good preconditioners for problems arising from (1.1).

We want to show in the present work that efficient iterative methods can be developed for this class of problems and that they make possible the solution of large problems with both accuracy and efficiency. We shall also insist on the fact that these methods should be entirely automatic and free of user dependent parameters.

© The Author(s), under exclusive license to Springer Nature Switzerland AG 2022
J. Deteix et al., *Numerical Methods for Mixed Finite Element Problems*,
Lecture Notes in Mathematics 2318, https://doi.org/10.1007/978-3-031-12616-1_1

We also want to make clear that our numerical results should be taken as examples and that we do not claim that they are optimal. Our hope is that they could be a starting point for further research.

Here is therefore our plan.

- Chapter 1 rapidly recalls the classical theory of mixed problems, including Augmented Lagrangian methods and their matricial form.
- Chapter 2 presents some classical iterative methods and describes the preconditioner which will be central to our development. We come back to augmented Lagrangian and a mixed form for penalty methods.
- Chapter 3 is devoted to numerical examples. The first one will be the approximation of a Dirichlet problem with Raviart-Thomas elements [20]. This is a simple case which will however permit to consider the fundamental issues. We shall see how an Augmented Lagrangian method enables us to circumvent the fact that we do not have coercivity on the whole space.

  We shall thereafter consider incompressible elasticity in solid mechanics, first in the linear case and then for a non linear Mooney-Rivlin model.

  In all those problems, the space of multipliers is $L^2(\Omega)$ and can therefore be identified with its dual. We also present some ideas for the solution of the Navier-Stokes equations. In those problems, with the discrete spaces that we employ, we shall not be able to use a real augmented Lagrangian. However, we shall consider a regularised formulation which will accelerate the convergence of our iterations.
- Chapter 4: We consider contact problems. In this case, the space of multipliers is not identified to its dual. We shall present some ideas for which we do not have numerical results but which we think could be research avenues for the future. In particular, we present ideas about discrete Steklov-Poincaré operators. Numerical results will be presented in the more classical formulation where the duality product is approximated by the L^2 scalar product.
- Chapter 5: Finally, we shall consider a case, arising from contact mechanics between incompressible bodies, in which we have two different types of constraints. This will lead us to modify accordingly our preconditioners.

Chapter 2
Mixed Problems

2.1 Some Reminders About Mixed Problems

Basically, mixed problems arise from the simple problem of minimising a quadratic functional under linear constraints. Let then V a function space and a bilinear form defining an operator A from V into V'.

$$a(u, v) = \langle Au, v \rangle_{V' \times V} \quad \forall u, v \in V$$

This bilinear form should be continuous, that is

$$a(u, v) \leq \|a\| \, \|u\|_V \, \|v\|_V \tag{2.1}$$

where $\| \cdot \|_V$ is the norm of V and $\|a\|$ is the norm of $a(\cdot, \cdot)$. In the same way, consider another function space Q, a bilinear form on $V \times Q$ defining an operator B from V into Q',

$$b(v, q) = \langle Bv, q \rangle \quad \forall v \in V, \forall q \in Q. \tag{2.2}$$

We also suppose this bilinear form to be continuous and thus

$$b(v.q) \leq \|b\| \, \|v\|_V \, \|q\|_Q \tag{2.3}$$

where $\|b\|$ is the norm of $b(\cdot, \cdot)$. Defining the functional

$$F(v) := \frac{1}{2} a(v, v) - \langle f, v \rangle \tag{2.4}$$

© The Author(s), under exclusive license to Springer Nature Switzerland AG 2022
J. Deteix et al., *Numerical Methods for Mixed Finite Element Problems*,
Lecture Notes in Mathematics 2318, https://doi.org/10.1007/978-3-031-12616-1_2

we then want to solve the constrained problem,

$$\inf_{Bv=g} F(v).$$ (2.5)

2.1.1 The Saddle Point Formulation

Problem (2.9) can be classically transformed into the saddle-point problem,

$$\inf_{v \in V} \sup_{q \in Q} \frac{1}{2} a(v, v) - b(v, q) - \langle f, v \rangle + \langle g, q \rangle,$$ (2.6)

for which the optimality system is

$$\begin{cases} a(u, v) + b(v, p) = \langle f, v \rangle & \forall v \in V \\ b(u, q) = (g, q)_Q & \forall q \in Q, \end{cases}$$ (2.7)

2.1.2 Existence of a Solution

For problem (2.7) to have a solution, it is clearly necessary that there exists some u_g satisfying $Bu_g = g$. Moreover, the lifting from g to u_g should be continuous. This is classically equivalent [20] to the inf-sup condition

$$\inf_{q \in Q} \sup_{v \in V} \frac{b(u, q)}{\|u\|_V \|q\|_Q} \geq k_0$$ (2.8)

a condition omnipresent in the following. Solving (2.5) is then equivalent to finding $u_0 \in Ker B$ such that,

$$a(u_0, v_0) = (f, v_0) - a(u_g, v_0) \quad \forall v_0 \in Ker\ B.$$ (2.9)

We shall refer to (2.9) as the *primal problem*. We therefore suppose, as in [20] that the bilinear form $a(\cdot, \cdot)$ is coercive on Ker B, that is there exists a constant α_0 such that,

$$a(v_0, v_0) \geq \alpha_0 \|v_0\|_V^2 \quad \forall v_0 \in Ker\ B$$ (2.10)

Remark 2.1 (Coercivity) Unless there is a simple way to build a basis of Ker B or a simple projection operator on Ker B coercivity on Ker B is not suitable for numerical computations. In our algorithms, we shall need, in most cases, coercivity on the whole of V.

$$a(v, v) \geq \alpha \|v\|_V^2 \quad \forall v \in V. \tag{2.11}$$

It is always possible to get this by changing $a(u, v)$ (see [43, 46, 79]) into

$$\tilde{a}(u, v) = a(u, v) + \alpha(Bu, Bv)_{Q'}.$$

Such a change will have different consequences in the development of our algorithms and will lead to augmented and regularised Lagrangians which will be considered in detail in Sect. 2.3   ∎

Remark 2.2 It should also be noted that (2.11) and (2.1) imply that $a(v, v)$ is a norm on v, equivalent to the standard norm.   ∎

2.1.3 Dual Problem

Problem (2.6) has the general form,

$$\inf_{v \in V} \sup_{q \in Q} L(v, q),$$

where $L(v, q)$ is a convex-concave functional on $V \times Q$. If one first eliminates q by computing

$$J(v) = \sup_{q \in Q} L(v, q),$$

one falls back on the original problem, the *primal problem*. Reversing the order of operations, (this cannot always be done, but no problems arise in the examples we present) and eliminating v from $L(v, q)$ by defining

$$D(q) := \inf_{v \in V} L(v, q)$$

leads to the dual problem

$$\sup_{q \in Q} D(q).$$

The discrete form of the dual problem and the associated Schur's complement will have an important role in the algorithms which we shall introduce.

2.1.4   A More General Case: A Regular Perturbation

We will also have to consider a more general form of problem (2.7). Let us suppose that we have a bilinear form $c(p, q)$ on Q. We also suppose that it is coercive on Q, that is

$$c(q, q) \geq \gamma |q|_Q^2$$

we thus consider the problem

$$\begin{cases} a(u, v) + b(v, p) = \langle f, v \rangle & \forall v \in V \\ b(u, q) - c(p, q) = (g, q)_Q & \forall q \in Q, \end{cases} \quad (2.12)$$

It is now elementary to show [20] that we have existence of a solution and

$$\alpha \|u\|_V^2 + \gamma \|p\|_Q^2 \leq \frac{1}{\alpha} \|f\|_{V'}^2 + \frac{1}{\gamma} \|g\|_{Q'}^2 \quad (2.13)$$

In practice we shall consider the perturbed problem defined with

$$c(p, q) = \epsilon(p, q)_Q.$$

The bound (2.13) explodes for γ small which can be a problem. If we still have the inf-sup condition and the coercivity on the kernel then we can have a bound on the solution independent of ϵ . We refer to [20] for a proof and the analysis of some more general cases.

Remark 2.3 In this case, the perturbed problem can be seen as a penalty form for the unperturbed problem. We shall come back to this in Sect. 2.3.4. It is then easy to see that we have an $O(\epsilon)$ bound for the difference between the solution of the penalised problem and the standard one. The bound depends on the coercivity constant and the inf-sup constant.   ∎

2.1.5   The Case $b(v, q) = (\mathcal{B}v, q)_Q$

The above presentation is abstract and general. We now consider a special case which will be central to the examples that we shall present later. Indeed, in many cases, it will more suitable to define the problem through an operator $\mathcal{B}$ from V into Q. We suppose that on Q, we have a scalar product defining a Ritz operator $\mathcal{R}$ from Q into Q'

$$(p, q)_Q = \langle \mathcal{R}p, q \rangle_{Q' \times Q} = (\mathcal{R}p, \mathcal{R}q)_{Q'}$$

We then define the bilinear form

$$b(v, q) = (\mathcal{B}v, q)_Q = \langle Bv, q \rangle_{Q' \times Q}$$

referring to (2.2) we then have,

$$B = \mathcal{R}\mathcal{B} \tag{2.14}$$

and

$$(Bu, q)_{Q' \times Q} = (\mathcal{B}u, q)_Q = \langle Bu, \mathcal{R}q \rangle_{Q \times Q'} \tag{2.15}$$

The constraint that we want to impose is then $\mathcal{B}u = g$ with $g \in Q$ or equivalently $Bu = g' = \mathcal{R}g$. In the case $Q = Q'$ and $\mathcal{R} = I$, the operators B and $\mathcal{B}$ coincide. This will not be the case for the problems which we shall consider in Chap. 5. But even in the case $Q = Q'$, in the discrete formulation the distinction between B and $\mathcal{B}$ will have to be taken into account.

2.2   The Discrete Problem

Our interest will be in solving discretisations of Problem (2.7). We thus suppose that we have subspaces $V_h \subset V$ and $Q_h \subset Q$. We shall work in the framework of Sect. 2.1.5 and we thus have,

$$b(v_h, q_h) = (\mathcal{B}v_h, q_h)_Q = (P_{Q_h}\mathcal{B}v_h, q_h)_Q \tag{2.16}$$

and we can define

$$\mathcal{B}_h = P_{Q_h}\mathcal{B}$$

The scalar product $(\cdot, \cdot)_Q$ defines an operator $\mathcal{R}_h$ from Q_h onto Q'_h and we can introduce

$$B_h = \mathcal{R}_h\mathcal{B}_h$$

We want to solve

$$\begin{cases} a(u_h, v_h) + b(v_h, p_h) = \langle f, v_h \rangle & \forall v_h \in V_h \\ b(u_h, q_h) = (g_h, q_h)_Q, & \forall q_h \in Q \end{cases} \tag{2.17}$$

We denote g_h the projection of g onto Q_h. The second equation of (2.17) can be read as $\mathcal{B}_h u_h = g_h$ or equivalently $B_h u = \mathcal{R}_h g_h$. Unless $\mathcal{B}u_h \in Q_h$, this condition is weaker than $\mathcal{B}u_h = g$ which for almost all cases will lead to bad convergence or even 'locking', that is a null solution.

For example, we shall meet later the divergence operator div (which acts on a space of vector valued functions which we shall then denote by $\underline{u}$). When we take for Q_h a space of piecewise constant functions, $\mathrm{div}_h\, \underline{u}_h$ is a local average of $\mathrm{div}\, \underline{u}_h$.

2.2.1 Error Estimates

We recall here, in its simplest form the theory developed in [20]. This will be sufficient for our needs. In the proof of existence of Sect. 2.1.2, we relied on two conditions: the coercivity on the kernel (2.10) and the inf-sup condition (2.8). We introduce their discrete counterpart. We thus suppose

$$\exists \alpha_0^h > 0 \text{ such that } a(v_0^h, v_0^h) \geq \alpha_0^h \|v_0^h\|_V^2 \qquad \forall\, v_0^h \in \mathrm{Ker}\, B_h, \tag{2.18}$$

or by the even simpler *global ellipticity*

$$\exists \alpha_h > 0 \text{ such that } a(v_h, v_h) \geq \alpha_h \|v_h\|_V^2 \qquad \forall\, v_h \in V_h.$$

We also write the discrete inf-sup condition,

$$\exists \beta_h > 0 \text{ such that } \sup_{v_h \in V_h} \frac{b(v_h, q_h)}{\|v_h\|_V} \geq \beta_h \|q_h\|_Q. \tag{2.19}$$

It will also be convenient, to define the **approximation errors**

$$E_u := \inf_{v_h \in V_h} \|u - v_h\|_V$$

$$E_p := \inf_{q_h \in Q_h} \|p - q_h\|_Q$$

We recall that approximation errors depend on the regularity of the solution. We refer to the classical work of [30] for an analysis of this dependence. Let $\|a\|$ and $\|b\|$ be the norms of $a(\cdot, \cdot)$ and of $b(\cdot, \cdot)$ as defined in (2.1) and (2.3). The following result is proved in [20].

Proposition 2.1 (The Basic Error Estimate) *Assume that V_h and Q_h verify (2.18) and (2.19). Let $f \in V'$ and $g \in Q'$. Assume that the continuous problem (2.17) has a solution (u, p) and let (u_h, p_h) be the unique solution of the discretised problem (2.17). If $a(\cdot, \cdot)$ is symmetric and positive semi-definite we have the estimates*

$$\|u_h - u\|_V \leq \left(\frac{2\|a\|}{\alpha_0^h} + \frac{2\|a\|^{1/2}\|b\|}{(\alpha_0^h)^{1/2}\beta_h} \right) E_u + \frac{\|b\|}{\alpha_0^h} E_p,$$

$$\|p_h - p\|_Q \leq \left(\frac{2\|a\|^{3/2}}{(\alpha_0^h)^{1/2}\beta_h} + \frac{\|a\|\,\|b\|}{\beta_h^2} \right) E_u + \frac{3\|a\|^{1/2}\|b\|}{(\alpha_0^h)^{1/2}\beta_h} E_p$$

Remark 2.4 (Dependance on h) In the previous result, we allowed, in principle, the constants β_h and α_1^h (or α_0^h) to depend on h. It is **obvious** (but still worth mentioning) that if there exist constants β_0 and α_0 such that $\beta_h \geq \beta_0$ and $\alpha_1^h \geq \alpha_0$ (or $\alpha_0^h \geq \alpha_0$) *for all h*, then the constants appearing in our estimates will be independent of h. Considering constants depending on h will be useful for contact problems (Sect. 5.2).   ∎

We can also see from this estimate that the approximation of p is more sensitive on the value of β_h than the approximation of u. One also sees that, paradoxically, improving E_u is better to counter this than improving E_p.

2.2.2 The Matricial Form of the Discrete Problem

To make explicit the numerical problem associated with (2.17) we need to introduce basis vectors ϕ_h^i for V_h and ψ_h^j for Q_h and the vectors of coefficients u, q which define u_h and q_h with respect to these bases.

$$u_h = \sum_i u_i \phi_h^i, \quad q_h = \sum_j q_j \psi_h^j. \tag{2.20}$$

Remark 2.5 (Notation) To avoid adding cumbersome notation, unless a real ambiguity would arise, we shall denote in the following by u and p either the unknowns of the continuous problem or the unknowns of the numerical problems arising from their discretisation.

We shall also denote by the same symbol the operators A and B of the continuous problem and the matrices associated to the discrete problems. As they are used in very different contexts, this should not induce confusion.   ∎

Denoting $\langle \cdot, \cdot \rangle$ the scalar product in $\mathbb{R}^n$, we can now define the matrices associated with the bases (2.20)

$$\langle A u, v \rangle = a(u_h, v_h).$$

We also have a matrix R associated with the scalar product in Q_h which represents the discrete Ritz operator

$$\langle Rp, q \rangle = (p_h, q_h)_{Q_h}.$$

The scalar product in Q'_h will be associated with R^{-1}.

$$\langle R^{-1} p, q \rangle = (p'_h, q'_h)_{Q'_h}.$$

We then have

$$\langle B u, q \rangle = b(u_h, q_h) = (\mathcal{B}_h u_h, q_h)_{Q_h}$$

and the operator $\mathcal{B}_h u_h$ can be asociated with the matrix

$$\mathcal{B} = R^{-1} B u \tag{2.21}$$

We summarise this in the following diagram.

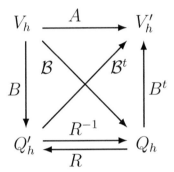

Remark 2.6 ($Q_h \neq Q'_h$) It is important to note that even in the case where $Q = Q'$ and $\mathcal{R} = I$, this is not the case in the discrete problem. The matrix R defined above is not the identity matrix.   ∎

Remark 2.7 (Change of Metric on Q_h)   As we shall see in Sect. 2.2.3, it will sometimes be useful in numerical applications to change the scalar product on Q_h. To distinguish this case we shall denote M_Q the matrix defining the scalar product

$$(p_h, q_h)_{Q_h} = \langle M_Q p, q \rangle.$$

The choice $M_Q = R$ is frequent but not optimal in many cases [48]. In Sect. 2.2.3 we shall discuss the choice $M_Q = M_S$ where M_S is some approximation of the discrete Schur complement.   ∎

We now consider the finite dimensional matricial problems associated to our mixed formulation, in fact the actual problem for which we want to build efficient solvers.

We first write problem (2.17) in matrix form,

$$\begin{pmatrix} A & B^t \\ B & 0 \end{pmatrix} \begin{pmatrix} u \\ p \end{pmatrix} = \begin{pmatrix} f \\ g \end{pmatrix} \tag{2.22}$$

Although this block matrix is a non singular matrix the numerical solution of (2.22) is not so simple. The main problem being that this matrix is indefinite. If one wanted to employ a direct solver, we might have to introduce a renumbering of the equations [28]. Moreover, our examples will come from the application of the mixed finite element methods [20]. Indeed, we shall focus on methods suitable for **large** systems arising from the discretisation of three-dimensional problems in mechanics which lead to large problems of the form (2.22). Such large systems are not suitable for direct solvers and require iterative methods. However, without preconditionning the system (acting on its eigenvalues), iterative methods, such as Krylov methods, are likely to diverge or stagnate on indefinite problems (see [86]). Therefore the key to obtain convergence will be the construction of a good preconditioner.

2.2.3 The Discrete Dual Problem: The Schur Complement

If we eliminate u from (2.22) we obtain the discrete form of the dual problem,

$$- B A^{-1} B^t p = g - B A^{-1} f. \tag{2.23}$$

The matrix $S = B A^{-1} B^t$ is often called the Schur complement. The system (2.23) is equivalent to the maximisation problem

$$\sup_{q} \frac{1}{2} \{ \langle -A^{-1} B^t q, B^t q \rangle + \langle g - B A^{-1} f, q \rangle \} \tag{2.24}$$

When building numerical methods, a crucial point will be the condition number of S. Indeed the condition number is an important measure for the behaviour (convergence) of iterative methods. Following Remark 2.7, we have on Q_h a metric defined by the matrix M_Q. The standard choice would be to take $M_Q = R$ where R is associated to the scalar product induced on Q_h by the scalar product of Q. In some cases, it will be convenient to change this scalar product, using a well chosen matrix M_S instead of R.

Remark 2.8 We write M_S to emphasise that this matrix should be chosen to approximate the Schur complement. ∎

We can change (2.23) into

$$M_S^{-1} B A^{-1} B^t p = M_S^{-1} (g - B A^{-1} f). \tag{2.25}$$

In Sect. 3.1, this will correspond to a **preconditioning**. The idea is that M_S should be an approximation of $BA^{-1}B^t$ improving the condition number of the resulting system.

To quantify this, we consider the eigenproblem

$$BA^{-1}B^t\phi_p = \lambda M_S\phi_p \tag{2.26}$$

Then the condition number for (2.25) can be defined as

$$K_h = \frac{\lambda_{max}}{\lambda_{min}}$$

We have written K_h to emphasise that we have a discretised problem. A nice property would be to have K_h independent of h. This would be the case if $\lambda_{max} \leq C$ and $\lambda_{min} \geq c$ with C and c independent of h. We then say that M_S and S are spectrally equivalent [9]. The ideal case would be $M_S = S$. Finding a good M_S means finding an easily computable approximation of S^{-1}.

Remark 2.9 (M_S Depends on A) We shall come back to the choice of M_S in Sect. 4.2 in particular for linear elasticity (or the Stokes problem). It is worth already remarking that M_S should be an approximation of BA^-B^t and therefore depends on A. In the simplest case, if A is changed into cA, one should change M_S into $\frac{1}{c}M_S$. ∎

Although the question of spectral equivalence depends on the problem at hand, we have a general case where it holds. We first recall that the minimum and maximum eigenvalues in (2.26) are the minimum and maximum values of the Rayleigh quotient,

$$RQ(q) = \frac{\langle A^{-1}B^t q, B^t q\rangle}{\langle M_S\, p, p\rangle}$$

Proposition 2.2 *We suppose that the matrix A defines a norm on V_h, equivalent to the standard norm.*

$$\alpha\|v_h\|^2 \leq \langle Av, v\rangle \leq \|A\|\,\|v_h\|^2.$$

We also suppose that we have the inf-sup condition (2.19). Then taking $M_S = R$, we have in (2.26)

$$\lambda_{min} \geq \frac{\beta_h^2}{\|A\|}, \quad \lambda_{max} \leq \frac{\|B\|^2}{\alpha}. \tag{2.27}$$

We recall that the dual norm associated to the norm defined by A is defined by A^{-1}. We can therefore write the Rayleigh quotient, as

$$\frac{\langle A^{-1}B^t q, B^t q \rangle}{\langle Rp, p \rangle} = \sup_v \frac{\langle v, B^t \rangle^2}{\langle Av, v \rangle \langle Rq, q \rangle}$$

The inf-sup condition yields the lower bound and the upper bound is direct.

This clearly also holds for another choice of M_S if we have

$$c_0 \langle Rp, q \rangle \le \langle M_S p, q \rangle \le c_1 \langle Rp, q \rangle \tag{2.28}$$

The bounds of (2.27) show that we have a condition number independent of the mesh size if $\beta_h \ge \beta_0 > 0$. This is an important property if we want to solve large-scale problems. However, if we consider M_S as a preconditioner, we shall also want to build it as a good approximation of S.

2.3 Augmented Lagrangian

Augmented Lagrangian is a popular method for the solution of mixed problems. We shall present it in some details, trying to show its potential but also its shortcomings.

2.3.1 Augmented or Regularised Lagrangians

If we employ the regularised bilinear form $\widetilde{a}(u, v)$ defined in (2.11) we can define an Augmented Lagrangian method.

$$\begin{cases} a(u, v) + \alpha(Bu, Bv)_{Q'} + b(v, p) = \langle f, v \rangle + \alpha(g', Bv)_{Q'} & \forall v \in V \\ b(u, q) = \langle g', q \rangle_{Q' \times Q}, & \forall q \in Q. \end{cases} \tag{2.29}$$

The extra term can be written differently using (2.15). We would then have,

$$\begin{cases} a(u, v) + \alpha(\mathcal{B}u, \mathcal{B}v)_Q + b(v, p) = \langle f, v \rangle + \alpha(g, \mathcal{B}v)_Q & \forall v \in V \\ b(u, q) = (g, q)_Q & \forall q \in Q \end{cases}$$

with $g = R^{-1}g'$. This does not change the solution of the problem. In the discrete problems, using an augmented Lagrangian is a classical way of accelerating some algorithms and cancelling errors associated with penalty methods. We shall have two different ways of implementing this idea.

The direct way to discretise (2.29) would be,

$$
\begin{cases}
a(u_h, v_h) + \alpha(\mathcal{B}u_h, \mathcal{B}v_h)_Q + b(v_h, p_h) \\
\qquad\qquad = \langle f, v_h \rangle + \alpha(g, \mathcal{B}v_h)_Q \qquad \forall v_h \in V_h \qquad (2.30) \\
b(u_h, q_h) = (g, q_h)_Q, \qquad\qquad\qquad\qquad \forall q_h \in Q_h.
\end{cases}
$$

This corresponds, for α large, to a penalty method imposing exactly $\mathcal{B}u_h = g$, in general a bad idea for many discrete problems. This will however be possible in some cases, such as the example of Sect. 4.1. In general, for small values of α, this formulation will be useful to regularise and improve some numerical solution algorithms. This will be the case in Sect. 4.2. We shall call this a **regularised formulation**.

To define a real discrete augmented Lagrangian, we suppose that Q_h has a scalar product defined by a metric M_Q. As in (2.21) we take $\mathcal{B}_h = P_{Q_h}\mathcal{B}$ which depends on M_Q. A real discrete Augmented Lagrangian would then be written as

$$
\begin{cases}
a(u_h, v_h) + \alpha(\mathcal{B}_h u_h, \mathcal{B}_h v_h)_{Q_h} + b(v_h, p_h) \\
\qquad\qquad = \langle f, v_h \rangle + \alpha(g_h, \mathcal{B}_h v_h)_{Q_h} \qquad \forall v_h \in V_h \\
b(u_h, q_h) = (g, q_h)_{Q_h}, \qquad\qquad\qquad\qquad \forall q_h \in Q_h, \\
\qquad\qquad\qquad\qquad\qquad\qquad\qquad\qquad\qquad\qquad\qquad (2.31)
\end{cases}
$$

which we can also write as

$$
\begin{cases}
a(u_h, v_h) + \alpha(B_h u_h, B_h v_h)_{Q'_h} + b(v_h, p_h) \\
\qquad\qquad = \langle f, v_h \rangle + \alpha(g_h, B_h v_h)_{Q_h \times Q'_h} \qquad \forall v_h \in V_h \\
b(u_h, q_h) = (g, q_h)_{Q_h}, \qquad\qquad\qquad\qquad \forall q_h \in Q_h. \\
\qquad\qquad\qquad\qquad\qquad\qquad\qquad\qquad\qquad\qquad\qquad (2.32)
\end{cases}
$$

This formulation depends on the choice of the metric M_Q on Q_h. The simplest case being when this scalar product is the scalar product in Q, that is $M_Q = R$. We could also consider a scalar product induced by a matrix M_S which we suppose equivalent to the usual scalar product. As in Sect. 2.2.3 M_S would be associated to a good approximation of the Schur complement.

The solution is unchanged by the extra terms and everything looks perfect. The question will rather be the actual implementation which will depend on the choice of finite element spaces and also on the **choice of the scalar product** on Q_h.

2.3.2 Discrete Augmented Lagrangian in Matrix Form

We now consider the matricial form of the augmented Lagrangian formulations. In the regularised form (2.30) we shall denote

$$\langle C\underline{u}, \underline{v} \rangle = (\mathcal{B}u_h, \mathcal{B}v_h)_Q$$

- For the regularised formulation (2.30) we have

$$\begin{pmatrix} A + \alpha C & B^t \\ B & 0 \end{pmatrix} \begin{pmatrix} u \\ p \end{pmatrix} = \begin{pmatrix} f + \alpha B^t g \\ g \end{pmatrix} \tag{2.33}$$

- From (2.32) we have

$$\begin{pmatrix} A + \alpha B^t M_S^{-1} B \ B^t \\ B & 0 \end{pmatrix} \begin{pmatrix} u \\ p \end{pmatrix} = \begin{pmatrix} f + \alpha B^t M_S^{-1} g \\ g \end{pmatrix} \tag{2.34}$$

Remark 2.10 (The Choice of M_S) The standard implementation is evidently to take $M_S = R$. If this is not the case, one should remark that in order to have (2.31) and (2.32) to be equivalent, we have to write $\mathcal{B}_h = M_S^{-1}B$ and not $R^{-1}B$ as in (2.21). ∎

Remark 2.11 (M_S^{-1} a Full Matrix?) Another important point is the presence of M_S^{-1} which is in general a full matrix. This makes (2.34) hardly usable unless M_S is diagonal or block diagonal and leads us to employ the regularised formulation. ∎

Remark 2.12 One could also write (2.34) writing the penalty term in mixed form, that is

$$\begin{pmatrix} A & B^t & B^t \\ B & -\epsilon M_S & 0 \\ B & 0 & 0 \end{pmatrix} \begin{pmatrix} u \\ \widehat{p} \\ p \end{pmatrix} = \begin{pmatrix} f \\ g \\ g \end{pmatrix}$$

The solution is of course $\widehat{p} = 0$. It is not clear if this form can be of some use. One might perhaps consider it in the context of Chap. 6. ∎

2.3.3 Augmented Lagrangian and the Condition Number of the Dual Problem

Using an augmented Lagrangian can be seen as a form of preconditioning for the dual problem. We have already considered in Sect. 2.2.3 the condition number of the dual problem (2.26), where M_S defines a scalar product on Q_h.

We had considered the eigenvalue problem,

$$BA^{-1}B^t\phi_p = \lambda M_S\phi_p \tag{2.35}$$

The eigenvalues of (2.35) are also the non zero eigenvalues of

$$A^{-1}B^t M_S^{-1} B\phi_u = \lambda\phi_u$$

with $\phi_u = A^{-1}B^t\phi_p$ which we can also write

$$A\phi_u = \frac{1}{\lambda} B^t M_S^{-1} B\phi_u$$

If instead of A we use $A + \alpha B^t M_S^{-1} B$ this becomes

$$(A + \alpha B^t M_S^{-1} B)\phi_u = \frac{1}{\lambda} B^t M_S^{-1} B\phi_u.$$

Denoting λ_α the corresponding eigenvalues, one easily sees that

$$\lambda_\alpha = \frac{\lambda}{1 + \lambda\,\alpha}$$

Denoting λ_M and λ_m the largest and the smallest eigenvalues, the condition number of the system is thus

$$K_\alpha = \frac{\lambda_M(1 + \alpha\,\lambda_m)}{\lambda_m(1 + \alpha\,\lambda_M)}$$

which converges to 1 when α increases. One sees that this holds for α large even if the initial condition number is bad. One also sees that improving $K = \lambda_M/\lambda_m$ also improves K_α for a given α. Augmented Lagrangian therefore seems to be the perfect solver. However, things are not so simple.

- The first problem is that the condition number of $A + \alpha\,B^t M_S^{-1} B$ worsens when α increases. As solving systems with this matrix is central to the algorithms that we shall introduce, we loose on our right hand what we gain with the left one. In practice, this means finding the correct balance between the conflicting effects.
- The other point is the computability of $B^t M_S^{-1} B$. The matrix M_S^{-1} could for example be a full matrix. Even if we approximate it by a diagonal matrix, the structure of the resulting matrix could be less manageable. This will be the case in the examples of Sect. 4.2.
- When the real augmented Lagrangian cannot be employed, the regularised formulation (2.33) might have a positive effect. However, the solution is perturbed and only small values of α will be admissible.

2.3.4 Augmented Lagrangian: An Iterated Penalty

Augmented Lagrangian formulations are often used as an iterative process, **using a penalty method** to correct a previous value of p. Given p^n, one solves

$$(A + \alpha B^t M_S^{-1} B) u^{n+1} + B^t p^n = f + \alpha B^t M_S^{-1} g$$

$$p^{n+1} = p^n + \alpha M_S^{-1} B u^n$$

This could also be written as, writing $\epsilon = 1/\alpha$

$$\begin{pmatrix} A & B^t \\ B & -\epsilon M_S \end{pmatrix} \begin{pmatrix} \delta u \\ \delta p \end{pmatrix} = \begin{pmatrix} f - A\,u^n - B^t p^n \\ g - Bu^n \end{pmatrix} = \begin{pmatrix} r_u^n \\ r_p^n \end{pmatrix} \tag{2.36}$$

and

$$p^{n+1} = p^n + \delta p, \quad u^{n+1} = u^n + \delta u.$$

System (2.36) is in fact a penalty method.

Remark 2.13 Problem (2.36) is of the perturbed form discussed in Sect. 2.1.4. Its solution will rely on the same iterative procedures as the unperturbed problem which we discuss in the next chapter. The question is whether there is some advantage in doing so.

- It permits to use an augmented Lagrangian without handling the matrix $B^t M_S^{-1} B$.
- The price is that it is an iterative process.

We must note that the Schur complement now becomes

$$S = B\,A^{-1}\,B^t + \epsilon M_S.$$

This makes the condition number of the perturbed problem better than that of the standard one which makes one hope of a better convergence for the dual variable.

One also sees that if M_S is an approximation of S one should rather use now,

$$M_S^\epsilon = (1 + \epsilon) M_S$$

This means that there will be an optimum value of $\epsilon = 1/\alpha$. Taking a larger ϵ for a better convergence enters in conflict with the paradigm of the augmented Lagrangian which would need α large and thus ϵ small. We shall present a numerical example in Remark 4.11.

■

Remark 2.14 (Correcting the Regularised Form) One could use a similar idea to eliminate the perturbation introduced by formulation (2.33), writing

$$
\begin{pmatrix} A + \alpha C & B^t \\ B & 0 \end{pmatrix} \begin{pmatrix} \delta u \\ \delta p \end{pmatrix} = \begin{pmatrix} f - A\,u^n - B^t p^n \\ g - B u^n \end{pmatrix} = \begin{pmatrix} r_u^n \\ r_p^n \end{pmatrix} \tag{2.37}
$$

again to price of an iteration.   ∎

Chapter 3
Iterative Solvers for Mixed Problems

We now come to our main issue, the numerical solution of problems (2.22), (2.33) and (2.34) which are indefinite problems, although with a well defined structure.

$$\mathcal{A} = \begin{pmatrix} A & B^t \\ B & 0 \end{pmatrix}.$$ (3.1)

Contact problems will also bring us to consider a more general non symmetric system

$$\mathcal{A} = \begin{pmatrix} A & B_1^t \\ B_2 & 0 \end{pmatrix}.$$ (3.2)

We intend to solve large problems and iterative methods will be essential. We shall thus first recall some classical iterative methods and discuss their adequacy to the problems that we consider. From there we introduce a general procedure to obtain preconditioners using a factorisation of matrices (3.1) or (3.2).

3.1 Classical Iterative Methods

Iterative methods is a topic in itself and has been the subject of many books and research articles. For the basic notions, one may consult [47, 50, 86] but this is clearly not exhaustive. Moreover the field is evolving and new ideas appear constantly. Our presentation will then be necessarily sketchy and will be restricted to the points directly relevant with our needs.

© The Author(s), under exclusive license to Springer Nature Switzerland AG 2022
J. Deteix et al., *Numerical Methods for Mixed Finite Element Problems*,
Lecture Notes in Mathematics 2318, https://doi.org/10.1007/978-3-031-12616-1_3

3.1.1 Some General Points

Linear Algebra and Optimisation

When considering iterative method, one may view things from at least two different points of view:

- linear algebra methods,
- optimisation methods.

These are of course intersecting and each one can bring useful information.

When dealing with systems associated to matrix (3.1), from the linear algebra perspective, we have an indefinite problem and from the optimisation perspective, we have a saddle point problem.

- We have positive eigenvalues associated with the problem in u and the matrix A is often symmetric positive definite defining a minimisation problem.
- On the other hand, we have negative eigenvalues associated to the problem in p, that is the dual problem (2.24) which is a maximisation problem.

This induces a challenge for iterative methods which have to deal with conflicting goals.

Norms

The linear systems which we consider arise from the discretisation of partial differential equations, and are therefore special. It is also useful to see if the iteration considered would make sense in the infinite dimensional case. Ideally, such an iterative method would have convergence properties independent of the mesh size.

Considering system (3.1), we have a very special block structure and variables u and p which represent functions $u_h \in V_h$ and $p_h \in Q_h$ that have norms which are not the standard norm of $\mathbb{R}^n$.

- This is an important point: we have a problem in **two** variables which belong to spaces with different norms.

In Remark 2.7 we introduced a matrix M_Q associated with a norm in Q_h, by the same reasoning we associate a matrix M_V for a norm in V_h. We thus have matrices M_V and M_Q associated with these norms

$$\langle M_V u, v \rangle = (u_h, v_h)_{V_h}, \quad \langle M_Q p, q \rangle = (p_h, q_h)_{Q_h}.$$

This also means that residuals must be read in a space with the dual norms M_V^{-1} and M_Q^{-1} and this will have an incidence on the construction of iterative methods.

Krylov Subspace

One should recall that classical iterative methods are based on the Krylov subspace,

$$\mathcal{K}r(A, b) = \text{span}(b, Ab, A^2 b, \ldots, A^{r-1} b),$$

looking for an approximate solution in this space. This is made possible by building an orthogonal basis. If the matrix is symmetric one uses the Lanczos method [60] to build orthogonal vectors. The important point is that symmetry allows to store a small and fixed number of vectors. This is the case in the conjugate gradient method and in the Minres [51, 74, 75, 83] algorithm.

- When A is symmetric positive definite (SPD) it defines a norm and one can look for a solution in $\mathcal{K}r(A, b)$ minimising $\|x - A^{-1} b\|_A^2 = \|Ax - b\|_{A^{-1}}^2$. This is the conjugate gradient method.
- When A is not positive definite, it does not define a norm. One then must choose a norm M and minimise $\|Ax - b\|_{M^{-1}}^2$. This is the Minres algorithm.

When the matrix is not symmetric, the Arnoldi process [7] can be used to build an orthogonal basis. This yields the GMRES algorithm [81] and related methods.

Preconditioning

We want to solve the problem $Ku = b$ and we suppose that we have a preconditioner P in the sense that P^{-1} is an approximation of K^{-1} (see [71, 88] for more details on the preconditioner techniques).

- For $K = \mathcal{A}$ we shall consider a class of preconditioner in Sect. 3.2.
- For the subproblems in u that is $(K = A)$, we shall rely on standard techniques such as those available in the Petsc package of Argonne Laboratory [11]. For example, we could employ as a preconditioner an Algebraic Multigrid method, a SSOR method or an incomplete factorisation. But this is a choice and using another technique is clearly permitted, the only criterion would be efficiency.

We shall also describe a special multigrid method in Sect. 3.1.4. The important point is that we want to solve large scale three-dimensional problems and that preconditioning will be a key issue.

We refer to [45] for a presentation and comparison of iterative methods and references.

Remark 3.1 (Preconditioning and Acceleration)   A preconditioner should ideally be a converging iterative method in itself. We can then consider that the preconditioned iterative method accelerates the preconditioner as well as the preconditioner accelerating the iterative method. In our development, we shall indeed try to build preconditioners which are convergent by themselves which may be hopefully accelerated by some other classical iteration.   ■

Given a preconditioner P and some iterative method IM we shall denote (P-IM) the method IM preconditioned by P.

Before considering the indefinite systems (3.1) or (3.2), we shall consider the subproblem associated with the matrix A which in many applications will be symmetric and positive definite. Here we are on safe ground.

3.1.2 The Preconditioned Conjugate Gradient Method

We first consider the case where A is a positive definite matrix. Solving $A u = b$ is equivalent to minimising a functional of the form (2.4). This is a convex functional and to minimise it, the standard procedure is a descent method. A simple preconditioned gradient method would be

$$u_{i+1} = u_i + \alpha_i P^{-1}(A u_i - b)$$

The residual $r = b - A u$ is then modified by

$$r_{i+1} = r_i - \alpha A P^{-1} r_i.$$

One then minimises in α

$$\|r_{i+1}\|^2_{A^{-1}}$$

which yields, denoting $z_i = P^{-1} r_i$

$$\alpha = \frac{\langle P^{-1} r_i, r_i \rangle}{\langle P^{-1} r_i, A P^{-1} r_i \rangle} = \frac{\langle z_i, r_i \rangle}{\langle z_i, A z_i \rangle}.$$

We must remark that we have two **different** notions.

- A norm on the residual, here defined by A^{-1}.
- A preconditioner P, that is an approximate solver.

This can be improved by the classical preconditioned conjugate gradient method (CG) [52, 81]. To use this we must have a symmetric positive definite $P = P^{1/2} P^{1/2}$. In fact, the method is equivalent to applying the standard conjugate gradient method to the system

$$P^{-1/2} A P^{-1/2} y = P^{-1/2} b. \tag{3.3}$$

which is equivalent to the minimisation problem

$$\inf_{v} \frac{1}{2}(A v, v)_{P^{-1}} - (b, v)_{P^{-1}} \tag{3.4}$$

with $u = P^{-1/2}y$. The preconditioning can thus be seen as a change of metric on the Euclidean space.

Remark 3.2 (Norm and Preconditioner)   If we come back to the idea that the residual is measured in the norm defined by A^{-1} one would now have this replaced by $P^{1/2}A^{-1}P^{1/2}$. With respect to the case where the standard $\mathbb{R}^n$ norm would be employed, we thus have two changes, one induced by A^{-1} and one by the preconditioner. This is special to the symmetric positive definite case and will not be directly transposed to the indefinite case where the two notions will have to be considered in two different ways.   ∎

Remark 3.3 (Convergence)   The principle of a conjugate gradient method is to keep the residual at every step orthogonal to all the previous directions in the norm defined by A^{-1}. This is done implicitly and does not require extra computation.

It can therefore be seen as a direct method as it will provide the solution of a problem in $\mathbb{R}^n$ in n steps. This would hold in exact arithmetic. Another nice property is superlinear convergence [87].   ∎

To be complete, we present explicitly the preconditioned conjugate gradient method

Algorithm 3.1 *P-CG algorithm*

1: Initialization

- *Let u_0 the initial value.*
- $r_0 = b - Au_0$
- $z_0 = P^{-1}r_0$
- $p_0 = z_0$
- $i = 0$

2: **while** *criterion $>$ tolerance* **do**

- $\alpha_i = \dfrac{r_i \cdot z_i}{p_i \cdot Ap_i}$
- $u_{i+1} = u_i + \alpha_i p_i$
- $r_{i+1} = r_i - \alpha_i Ap_i$
- $z_{i+1} = P^{-1}r_{i+1}$
- $\beta_i = \dfrac{z_{i+1} \cdot r_{i+1}}{z_i \cdot r_i}$
- $p_{i+1} = z_{i+1} + \beta_i p_i$
- $i = i + 1$

end while   ∎

3.1.3   Constrained Problems: Projected Gradient and Variants

When considering contact problems in Chap. 5 we shall have to solve a minimisation problem with a positivity constraint. The method that we shall use is a special case of a more general method.

Equality Constraints: The Projected Gradient Method

The projected gradient method is a very classical method for constrained optimisation [18]. In the simpler case, let us suppose that the solution of a minimisation problem must satisfy a linear constraint,

$$Bu = g \tag{3.5}$$

If we have an initial value u_g satisfying (3.5) and if we have a **simple** way of projecting on Ker B, we can solve the problem,

$$\inf_{v_0 \in Ker B} \frac{1}{2} \langle A \, v_0 - A u_g, v_0 \rangle - \langle f, v_0 \rangle.$$

This is in fact what we did to prove existence of the mixed problem. We shall meet this procedure in Sect. 6.2. We can then apply the conjugate gradient method provided the gradient is projected on Ker B. We shall also consider this method in Sect. 4.3.

Inequality Constraints

We now consider a set of inequality constraints,

$$B_i \, u \leq g_i \quad 1 \leq i \leq m$$

Let u be the solution, we then split the constraints in two sets.

- Active constraints for which we have $B_i u = g_i$.
- Inactive constraints for which $B_i u < g_i$.

For more details on this strategy, we refer to [16, 17, 56, 57, 72]. If we know which constraints are active, the problem reduces to the case of equality constraints. In practice, this will be done iteratively.

- Given an initial guess for the active set, use an iterative method to solve the equality problem.
- Monitor the possible changes of active constraints. If there is a change, restart the iteration with the new set.

The monitoring implies a change of active constraints if one has one of two conditions.

- The solution is modified such that an inactive constraint is violated, one then projects the solution on the constraint and restarts the iteration, now making this constraint active.
- On an active constraint, the iteration creates a descent direction that would bring the solution to the unconstrained region. This constraint is then made inactive.

This method is specially attractive in the following case.

Positivity Constraints

An important special case is when the solution must satisfy $u_j \geq 0$. The constraint is active if $u_j = 0$ and inactive if $u_j > 0$. The projection on the active set is readily computed by putting inactive values to zero (see [3, 54, 55]). The gradient (or more generally the descent direction) is also easily projected. We shall meet this case in contact problems (Sect. 5.2.2).

Convex Constraints

To complete this section, we give a hint on how this can be extended to a convex constraint [18]. We consider as in Fig. 3.1 a point u_0 on the boundary of the convex set C. If the constraint is active, the gradient (or some descent direction) z is pointing to the exterior of C. We can then project the gradient on the tangent to C (in red in Fig. 3.1), search for an optimal point u^* on the tangent as we now have a linear constraint, and then project the result on C to obtain u_1. This will converge if the curvature of the boundary of C is not too large.

Fig. 3.1 Projected gradient for convex constraint

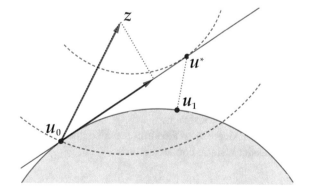

3.1.4 Hierarchical Basis and Multigrid Preconditioning

There is a whole literature on preconditioning for standard finite element problems. It would be out of scope to present it here. Let us just say, for example, that incomplete factorisation is often very efficient. A classical combination is also to use a SSOR method to precondition a CG method.

Multigrid methods [27, 70, 81] yield powerful preconditioning. On structured meshes, they can be built very efficiently. On a general mesh, algebraic multigrid methods (AMG) are also available. In our computational examples, we shall often use a multigrid method based on a hierarchical basis. This was described in [1, 38]. The principle is simple and we use it in the case of a second degree finite element approximation in which we represent these elements by a hierarchical basis.

- The basis function associated to vertices is the standard first degree basis function.
- The basis function on edges is associated, not to a nodal value but to a correction to the piecewise linear approximation.

When this is done, the matrix A can be subdivided in submatrices

$$A = \begin{pmatrix} A_{11} & A_{12} \\ A_{21} & A_{22} \end{pmatrix}$$

The preconditioning consists in solving with a block Gauss-Seidel iteration for the piecewise linear part (A_{11}) and the piecewise quadratic correction on the edges (A_{22}).

- It must be noted that for a typical three dimensional mesh, matrix A_{11} is about eight times smaller than the global matrix A. This makes it possible to use a direct method to solve the associated problem.
- This being a multigrid method, it can also be completed by applying an Algebraic Multigrid (AMG) method to the problem in A_{11}.
- The matrix A_{22} has a small condition number independent of the mesh size [89] and a SSOR method is quite appropriate.

In our examples, this preconditioner will be applied to the rigidity matrix of elasticity problems. As we shall see, it is quite efficient for large problems.

The idea of hierarchical basis could be employed for other discretisations. In some finite element approximations, degrees of freedom are added on the faces to better satisfy a constraint.

Internal degrees of freedom are also amenable to this technique in a variant of the classical 'static condensation'.

3.1.5 Conjugate Residuals, Minres, Gmres and the Generalised Conjugate Residual Algorithm

We now consider the more general case of a system

$$Kx = b$$

having in mind $K = \mathcal{A}$ as in (3.1) or more generally (3.2).

When the matrix K is symmetric but not positive definite, the idea of keeping a residual orthogonal to all the previous ones as in the conjugate gradient method is still possible. However we must now choose a norm defining this orthogonality as K^{-1} can no longer be used. This chosen norm will also be the norm in which we shall minimise the residual. The simplest thing is to use the euclidean norm of $\mathbb{R}^n$. This is the Conjugate Residuals (CR) method for symmetric systems presented in [65].

Considering Remark 3.1.1 one should rather introduce some norm which is better adapted to our problem. Contrarily to the case of the Conjugate gradient method, making the residuals orthogonal in some norm M^{-1} requires some extra computation. Some examples can be found in [67] and also in [64] where a general theory is developed. If $r = b - Kx$, $z = M^{-1}r$ and $T = Kr$ the basic idea is thus to minimize

$$\|r - \alpha T\|_{M^{-1}}^2$$

which yields

$$\alpha = \frac{\langle r, M^{-1}T \rangle}{\langle T, M^{-1}T \rangle}$$

This can be written in many equivalent ways. We refer for example to [41, 49] for a complete presentation. In the classical Minres algorithm, the Lanczos method is completed by Givens rotation to achieve orthogonality of residuals.

Remark 3.4 Although Minres is a well studied and popular method, we did not use it for three reasons.

- Preconditioning: the change of metric is often said to be a SPD preconditioning. We shall present in Sect. 3.2 preconditioners which are symmetric but not positive definite.
- Contact problems are constrained problems and one must have access to the part of the residual with respect to the multiplier. This was not possible to our knowledge with the standard Minres implementation or indeed with GMRES (see however [51]).
- Frictional contact leads to non symmetric problems.

■

Our choice was rather to consider a method allowing both a general preconditioning as a change of metric and non symmetrical matrix.

The Generalised Conjugate Residual Method

If the conjugate residual method introduced by Luenberger [65] seemed a natural starting point, our interest in solving non symmetrical problems arising from frictional contact led us to the use of the generalised conjugate residual (GCR) method ([36, 39], which can be seen as a variant of the flexible GMRES method [80]). Evidently more standard methods such as the Conjugate Gradient (CG) method can be preferred when their use is possible.

Since the GCR method does not rely on symmetry, which means that non symmetrical preconditioners can be employed. The price to pay is the storage of a stack of vectors. When using the GCR method, we have two possibilities to consider preconditioning.

- Left preconditioning: the system $Kx = b$ is replaced by the equivalent system $P^{-1}Kx = P^{-1}b$.
- Right preconditioner: the system is replaced by the system $KP^{-1}y = b$ with $x = P^{-1}y$.

Remark 3.5 (Change of Metric) We can introduce a metric M different from the standard euclidean metric on $\mathbb{R}^n$ in the left and right preconditioned GCR algorithm. The metric M is applied in the space of solutions while the dual metric M^{-1} is applied to the residuals. ∎

We thus have two possibilities while using the preconditioner P to accelerate a GCR method. We give here both left and right generic P-GCR using an arbitrary metric M.

The Left Preconditioning

We first present the left preconditioner. This is the standard procedure as presented to precondition the GMRES method in [81].

Algorithm 3.2 *Left preconditioned P-GCR algorithm with an arbitrary metric M*

1: Initialization

- $i = 0$
- *Let x_0 the initial value.*
- $r_0 = b - K x_0$
- $r_0^P = P^{-1}(r_0)$

2: **while** *criterion > tolerance* **do**

- $z_i = r_i^P$
- $T_i = K z_i$
- $z_i = P^{-1} T_i$
- *From z_i, using the **modified Gram-Schmidt (MGS)**, compute $z_i^\perp$ **orthonormal in the M-norm** to $[z_0^\perp, \cdots, z_{i-1}^\perp]$. Using the same transformation on T_i compute $T_i^\perp$ based on $[T_0^\perp, \cdots, T_{i-1}^\perp]$.*
- $\beta = (r_i^P, M z_i^\perp)$
- *Update*

$$
\begin{cases}
r_{i+1} = r_i - \beta T_i^\perp \\
r_{i+1}^P = r_i^P - \beta z_i \\
x_{i+1} = x_i + \beta z_i \\
i = i + 1
\end{cases}
$$

endwhile   ■

The Right Preconditioning

This is the method which was used in most of our numerical results.

Algorithm 3.3 *Right preconditioned $P-GCR$ algorithm with an arbitrary metric M^{-1}*

1: *Initialization*

- $i = 0$
- *Let x_0 the initial value.*
- $r_0 = b - K x_0$

2: **while** *criterion > tolerance* **do**

- $z_i = P^{-1} r_i$.
- $T_i = K z_i$
- *From T_i, using the MGS, compute $T_i^\perp$ **orthonormal in the M^{-1}-norm** to $[T_0^\perp, \cdots, T_{i-1}^\perp]$. Using the same transformation to z_i compute $z_i^\perp$ based on $[z_0^\perp, \cdots, z_{i-1}^\perp]$.*
- $\beta = \langle r, M^{-1} T_i \rangle$

- *Update*

$$\begin{cases} x_{i+1} = x_i + \beta z_i^{\perp} \\ r_{i+1} = r_i - \beta T_i^{\perp} \\ i = i + 1 \end{cases}$$

endwhile   ■

Remark 3.6 (Choice of Metric) If M^{-1} is the identity, that is the standard metric of $\mathbb{R}^n$, this reduce to the classical algorithm. This raise the question of the choice of M. To fix ideas, let us consider two cases frequently met.

When K and the preconditioner P are both SPD. We would then want to obtain a method equivalent to the preconditioned conjugate gradient method. As we have seen in Remark 3.2 the norm should then be defined using $P^{1/2}K^{-1}P^{1/2}$. If the preconditioner is good, this is close to identity. Thus $M = I$ is not a bad choice when we have a good precondioner.

For a symmetric system, if the preconditioner is SPD, we could take $M = P$. One could then read this as a variant of Minres.   ■

The Gram-Schmidt Algorithm

In the GCR methods described above, we want to make a vector T_i orthogonal to each vector in a stack in norm M or M^{-1}. We consider for example the case of M-orthogonality. If the vectors in the stack are orthonormal in norm M, the modified Gram-Schmidt method means computing,

Algorithm 3.4 *Modified Gram-Schmidt orthonormalisation*

1: Initialisation

- T_i *given*

2: for $j = i - 1, \ldots, 0$ do

- $s_j = \langle M T_i, T_j^{\perp} \rangle$,
- $T_i = T_i - s_j T_j^{\perp}$

endfor

- $T_i^{\perp} = T_i / \|T_i\|_M$

■

For the modified Gram-Schmidt method, one thus needs to compute $M T_i$ at every step as T_i is changed, which could be expensive. This can be avoided at the cost of storing in the stack both $T_j^{\perp}$ and $M T_j^{\perp}$.

GCR for Mixed Problems

We now come to our central concern: solving mixed problems. We now have an indefinite system. In that case, iterative methods will often diverge or stagnate [45]. For example, without preconditioning a conjugate gradient method applied to an indefinite system will diverge.

Remark 3.7 (We Have a Dream) As we have already stated, we are dealing with a problem in two variables. We dream of a method which would really take this fact into account. The preconditioner that we shall develop will. The GCR method is then not really satisfying even if it provides good results.

   Alas, nothing is perfect in our lower world.   ■

   To fix ideas, let us consider a direct application of a right preconditioned GCR algorithm. Let then $\mathcal{A}$ be as in (3.1) and $\underline{b} = (f, g) \in \mathbb{R}^n \times \mathbb{R}^m$. We are looking for $\underline{x} = (u, p)$ and we use the right preconditioned GCR as in Algorithm 3.3. As before we introduce an arbitrary metric. In this case the metric take into account the mixed nature of the problem and is characterised by two different matrices M_u and M_p giving a (M_u, M_p)-norm for (u, p).

$$\|(u, p)\|^2_{(M_u, M_p)} = \|u\|^2_{M_u} + \|p\|^2_{M_p} = (M_u u, u) + (M_p p, p)$$

Obviously, if both M_u and M_p are identities we have the usual right preconditionned GCR algorithm.

Algorithm 3.5 *Right preconditioned Mixed-P-GCR algorithm with arbitrary metric (M_u, M_p)*

 1: Initialization

 - *$i=0$;*
 - *Let $\underline{x}_0 = (u_0, p_0)$ the initial value.*
 - *$r_u = f - Au_0 - B^t p_0$, $r_p = g - Bu_0$ and $\underline{r} = (r_u, r_p)$*

 *2: **while** criterion > tolerance **do***

 - *From $\underline{r} = (r_u, r_p)$ the preconditionner yields $\underline{z}_i = (z_{iu}, z_{ip})$.*
 - *$T_{iu} = Az_{iu} + B^t z_{ip}$, $T_{ip} = B z_{iu}$*
 - *From $\underline{T}_i = (T_{iu}, T_{ip})$, using the MGS compute $\underline{T}_i^\perp$ orthonormal in (M_u^{-1}, M_p^{-1})-norm to $[\underline{T}_0^\perp, \cdots, \underline{T}_{i-1}^\perp]$. Using the same transformation to $\underline{z}_i$ compute $\underline{z}_i^\perp$ based on $[\underline{z}_0^\perp, \cdots, \underline{z}_{i-1}^\perp]$.*
 - *$\beta = \langle r_{iu}, M_u^{-1} T_{iu}^\perp \rangle + \langle r_{ip}, M_p^{-1} T_{ip}^\perp \rangle$*

- *Update*

$$\begin{cases} \underline{x}_{i+1} = \underline{x}_i + \beta \underline{z}_i^{\perp} \\ \underline{r} = \underline{r} - \beta \underline{T}_i^{\perp} \\ i = 1 + 1 \end{cases}$$

end while   ∎

Remark 3.8 In the above GCR algorithm, we could decompose β in two component

$$\beta = \beta_u + \beta_p \qquad \beta_u = \langle r_u, M_u T_{iu}^{\perp} \rangle, \quad \beta_p = \langle r_p, M_p T_{ip}^{\perp} \rangle$$

If the system is not properly preconditioned, β_u or β_p can become negative and $\beta = \beta_u + \beta_p$ can even become null, hence stagnation. This is the symptom of a bad preconditioning. Another issue is that giving an equal weight to u and p in the scalar product is not a priori the best choice, although the right choice is seldom clear. This fact is in reality common to any iterative method applied to a system of this form.   ∎

The key is therefore in the construction of good preconditioners or in building optimisation methods adapted to saddle-point problems. This will be our next point.

Remark 3.9 (The Case of Sect. 2.1.4: Perturbed Problems)   Whenever one wants to solve a problem of the form (2.12) one should make a few changes.

- In the initialisation phase, one should have $r_p = g - Bu - \epsilon Rp$
- After the preconditioner, one should have $T_p = Bz_u - \epsilon Rz_p$.

                                                                         ∎

3.2 Preconditioners for the Mixed Problem

3.2.1 Factorisation of the System

This section is centered on the development of an efficient preconditioner for the system

$$\mathcal{A} \begin{pmatrix} u \\ p \end{pmatrix} = \begin{pmatrix} f \\ g \end{pmatrix} \tag{3.6}$$

with $\mathcal{A}$ an indefinite matrix of the form (3.1) or more generally for the non symmetric case (3.2). Our first step toward a general solver for systems (3.6) will be to consider a block factorisation $\mathcal{LDU}$ of the block matrix $\mathcal{A}$

$$\mathcal{A} = \begin{bmatrix} I & 0 \\ B_1 A^{-1} & I \end{bmatrix} \begin{bmatrix} A & 0 \\ 0 & -S \end{bmatrix} \begin{bmatrix} I & A^{-1}B_2^t \\ 0 & I \end{bmatrix} := \mathcal{LDU} \tag{3.7}$$

where the Schur complement which we already considered in Sect. 2.2.3 $S = B_1 A^{-1} B_2^t$ is invertible, we then have,

$$(\mathcal{LDU})^{-1} = \begin{bmatrix} I & -A^{-1}B_2^t \\ 0 & I \end{bmatrix} \begin{bmatrix} A^{-1} & 0 \\ 0 & -S^{-1} \end{bmatrix} \begin{bmatrix} I & 0 \\ -B_1 A^{-1} & I \end{bmatrix} \tag{3.8}$$

If A is symmetric and $B_1 = B_2$ we have $\mathcal{U} = \mathcal{L}^T$ and the full block decomposition $\mathcal{F} \simeq \mathcal{LDL}^T$ is symmetric but indefinite. To simplify, we shall focus our presentation on the symmetric case as the extension to the non symmetric case is straightforward.

Remark 3.10 (Regular Perturbation)   We have considered in Sect. 2.1.4 the case of a regular perturbation where $Bu = g$ is replaced by $Bu - \epsilon Mp = g$. In that case S is replaced in the factorisation (3.7) by $S + \epsilon M$ ∎

Solving Using the Factorisation

Let (r_u, r_p) be the residual of the system (3.6) for (u_0, p_0)

$$r_u = f - Au_0 - B^t p_0 \qquad r_p = g - Bu_0.$$

Using the factorisation (3.8) to obtain $(u, p) = (u_0 + \delta u, p_0 + \delta p)$ leads to three subsystems, two with the matrix A and one with the matrix S.

Algorithm 3.6

$$\begin{cases} \delta u^* = A^{-1} r_u \\ \delta p = S^{-1}(B\delta u^* - r_p) \\ \delta u = A^{-1}(r_u - B^t \delta p) = \delta u^* - A^{-1} B^t \delta p. \end{cases} \tag{3.9}$$

∎

It must be noted that this is symmetric as it is based on (3.8). Particular attention must be given to the solution

$$S\,\delta p = BA^{-1}B^t \delta p = B\delta u^* - r_p. \tag{3.10}$$

With the exception of very special cases, S is a full matrix and it is not thinkable of building it explicitly. Even computing matrix-vector product (needed by any iterative solver) would require an exact solver for A which may become impossible for very large problems. Therefore, except in rare cases where systems in A can be solved extremely efficiently Algorithm 3.6 is practically unusable. Thus in most cases efficiency will lead us to iterative solvers for $\mathcal{A}$ which can avoid the need for an exact solver for S.

Nevertheless (3.9) suggests a way to precondition $\mathcal{A}$. To do so, we shall first write an approximate form of the factorisation (3.7).

Assuming that we have an 'easily invertible' approximation $\widetilde{A}$ of A and inserting it in (3.7) to replace A, we obtain

$$\widetilde{\mathcal{A}} = \begin{bmatrix} I & 0 \\ B\widetilde{A}^{-1} & I \end{bmatrix} \begin{bmatrix} \widetilde{A} & 0 \\ 0 & -\widetilde{S} \end{bmatrix} \begin{bmatrix} I & \widetilde{A}^{-1}B^t \\ 0 & I \end{bmatrix} := \mathcal{LDU} \tag{3.11}$$

where

$$\widetilde{S} = B\widetilde{A}^{-1}B^t \tag{3.12}$$

If we now have an approximation $\widehat{S}$ of $\widetilde{S}$, we can introduce

Algorithm 3.7 *A general mixed preconditioner (GMP)*

$$\begin{cases} z_u^* = \widetilde{A}^{-1}r_u \\ z_p = \widehat{S}^{-1}(Bz_u^* - r_p) \\ z_u = \widetilde{A}^{-1}(r_u - B^t z_p) = z_u^* - \widetilde{A}^{-1}B^t z_p. \end{cases} \tag{3.8}$$

∎

Using the factorisation (3.11), we may associate (3.8) with a matrix $\widehat{\mathcal{A}}$

$$\widehat{\mathcal{A}} = \begin{bmatrix} \widetilde{A} & B^t \\ B & B\widetilde{A}^{-1}B^t - \widehat{S} \end{bmatrix} \tag{3.9}$$

This suggests many variants for a preconditioner of $\mathcal{A}$. Their implementation relies on two essential tools.

1. An easily computable and fast approximate solver $\widetilde{A}^{-1}$. For this, we can rely on classical iterative methods such as the *Conjugate Gradient* method when the matrix A is symmetric and positive definite, the *GMRES* method or the *Generalised Conjugate Residual* method. Many other methods are available on general packages such as Petsc or Lapack. In all cases, a good preconditioner will be essential. We shall often rely on the multigrid method of Sect. 3.1.4.
2. An easily computable and fast solver $\widehat{S}$ to approximate $\widetilde{S}$.

This contains in particular the case $\widetilde{A} = A$ and therefore $\widetilde{S} = S$.

Remark 3.11 (A Solver in Itself)   Although we shall use Algorithm 3.7 as a preconditioner, it should be noted that it is also an iterative method in itself and that it can be employed as a solver. Using it as a preconditioner in an iterative method can be seen as a way of accelerating it (Remark 3.1).

3.2.2 Approximate Solvers for the Schur Complement and the Uzawa Algorithm

In the following, we consider as in (3.12) approximations of the Schur complement S or its approximate form $\widetilde{S}$ defined in (3.12).

When the solver in u is exact, we have $\widetilde{S} = S = BA^{-1}B^t$ and the preconditioner of Algorithm 3.7 reduces to using $\widehat{S}$ instead of S. Therefore, in any cases, $\widehat{S}$ is related to a Schur complement matrix. We will not explicitly construct a matrix $\widehat{S}$. Instead the equation $\widehat{S}v = r$ in (3.8) should be interpreted as an approximate solution of $\widetilde{S}v = r$. We thus describe how we can employ an iterative method to obtain a computable $\widehat{S}$. This iterative method can then be used with a more or less stringent desired precision or a limited number of iterations. We also need a preconditioner for this iteration, proceeding as follows

- We introduce an easily invertible approximation M_S of S, following the discussion Sect. 2.2.3
- We solve $\widetilde{S}z_p = r_p$ by a an iterative method using M_S as a preconditioner. In practice we restrict this step to a few (most often one) iteration.

Remark 3.12 (The Choice of M_S)   This is an important point as it can make the difference between an efficient method and a poor one. The choice evidently depends on the problem. We have considered in Sect. 2.2.3 and in particular in Proposition 2.2 how the choice for $M_S = R$ or M_S satisfying (2.28) would be adequate provided the inf-sup condition holds and $\langle Av, v \rangle$ is a norm on V_h, that are the conditions to have a well posed problem.

- For problems of incompressible materials which we shall consider in Sect. 4.2, the choice $M_S = M_0$ where we denote M_0 the 'mass matrix' associated to the L^2 scalar product in Q is a good choice. Indeed it was shown in [41] that this choice leads to an iteration count independent of the mesh size (See Proposition 2.2.3). In this case S is an operator of order 0.
- For contact problems, $M_S = M_0$ is less efficient as we do not have the inf-sup condition in L^2 and S is now of order 1. We shall discuss this in Sect. 5.1. We shall also show how a discrete Steklov-Poincaré operator could be built.
- One could also think of a BFGS (Broyden, Fletcher, Goldfarb, and Shannon) update [63] to improve M_S during the iterative process.
- One should also recall (Sect. 2.3.3) that the Augmented Lagrangian Method yields a preconditioning for the dual problem.

■

We now make explicit two iterative processes. As we noted above, they will be the same for the solution of $Sp = r_p$ or $\widetilde{S}p = r_p$. For the GCR method, we have the following algorithm.

Algorithm 3.8 *A right preconditioned GCR iterative solver for* $\widetilde{S}$

1: *Initialization*

  - *i=0;*
  - *Let z_u and r_{0p} be given values.*
  - $z_p = 0$

2: **while** *criterion > tolerance or maximum number of iterations* **do**

  - $zz_{ip} = M_{\widetilde{S}}^{-1} r_{ip}$
  - $zz_{iu} = -\widetilde{A}^{-1} B^t zz_{ip}$
  - $T_{ip} = B zz_{iu}.$
  - *Use the MGS procedure to obtain $T_{ip}^{\perp}$ orthonormal to the previous directions.*
    Apply the same transformation to zz_{ip} and zz_{iu} to get $zz_{ip}^{\perp}$ and $zz_{iu}^{\perp}$.
  - $\beta = (T_{ip}^{\perp}, r_{ip})$
  - *Update*

$$\begin{cases} z_p = z_p - \beta zz_{ip}^{\perp} \\ z_u = z_u - \beta zz_{iu}^{\perp} \\ i = i + 1 \end{cases}$$

  end while   ∎

One could also use the more general version of Algorithm 3.3 with a change of metric.

Remark 3.13 The computation of z_u is optional. It avoids an additional use of $\widetilde{A}^{-1}$ when this is included in the preconditioner 3.7 to the price of some additional work in the Gram-Schmidt process.   ∎

If $\widetilde{A}$ and M_S are symmetric, we can also use the simpler Conjugate Gradient method, which takes the following form

Algorithm 3.9 *A CG iterative solver for* $\widetilde{S}$

1: *Initialisation*

  - $i = 0$
  - z_u *and r_p, given values*

2: **while** *criterion > tolerance or maximum number of iterations* **do**

  - $zz_p = M_S^{-1} r_p$
  - *If $i > 0$* $\alpha = (zz_p, r_p)/(zz_p^0, r_p^0)$
  - $w_p = zz_p + \alpha w_p^0$
  - $zz_u = -\widetilde{A}^{-1} B^t w_p$
  - $T_p = B zz_u.$

- $\beta = (w_p, r_p)/(T_p, w_p)$
- $w_p^0 = w_p,\ zz_p^0 = zz_p,\ r_p^0 = r_p$
- *Update*

$$\begin{cases} z_p = z_p + \beta zz_p \\[4pt] z_u = z_u - \beta zz_u \\[4pt] r_p = r_p - \beta T_p. \\[4pt] i = i + 1 \end{cases}$$

end while   ∎

The Uzawa Algorithm

When $\widetilde{A} = A$ we recall that $\widetilde{S} = S$ and for sufficiently strict convergence criteria the last algorithm corresponds to a solver for S. Then when included in Algorithm 3.7 it yields solutions of (3.6) that is Algorithm 3.7 coincides with Algorithm 3.6. In [43], this was called the Uzawa algorithm, which we can summarise as formed of two steps illustrated in Fig. 3.2.

- Solve the unconstrained problem $Au = f$.
- Project this solution, in the norm defined by A, on the set $Bu = g$.

In [43] Uzawa's algorithm was presented in its simplest form depending on an arbitrary parameter β. The parameter β must then be chosen properly and depends on the spectrum of $BA^{-1}B^t$. This was studied in detail and convergence follows as in the classical analysis of gradient methods (an acceleration by a conjugate gradient method was also considered. Using this method implies that one has an efficient solver for A.

Many variants of this method are proposed in the literature (see [15, 81]). Several authors have studied numerically and in the theoretical framework of the variants of the method. Elman and Golub [40] proposed the Uzawa method called

Fig. 3.2 Illustration of the Uzawa algorithm

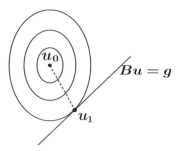

inexact whose theoretical analysis is made in [23]. Bai et al. [10] presented the so-called *parametrised* method. More recently, Ma and Zang in [66], dealt with the so-called *corrected* method. They have shown that it converges faster than the classical method and several of its variants under certain assumptions. However, their approach come up against the question of determining the optimal parameter.

3.2.3 The General Preconditioned Algorithm

We now come to the use of Algorithm 3.7: the General Mixed Preconditioner.

- We must first chose $\widetilde{A}^{-1}$. For this we rely on standard and well proven iterative methods. Whenever possible we precondition these methods by a multigrid procedure.
- We also need an approximate solver for the approximate Schur complement $\widehat{S}$.
- We also choose a norm N in which we minimise residuals. This will most of times be the euclidean norm but better choices are possible

If for $\widehat{S}$ we use Algorithm 3.9 or Algorithm 3.8 (i.e. we solve $\widetilde{S}$) then we use a limited number of iterations. Choosing $\widehat{S} = \widetilde{S}$ (i.e. fully converging Algorithm 3.9 or Algorithm 3.8) is a possibility, but in general it is not a good idea to solve too well something that you will throw away at the next iteration. We shall thus develop the case where only one iteration is done.

Remark 3.14 One should note that using a better $\widetilde{A}$ and more iterations for $\widetilde{S}$ is a direct way to make things better.

- If $\widetilde{A} = A$ we have the Uzawa algorithms or close variants.
- If $\widetilde{S}$ is solved exactly, we have a form of the projected gradient method.

∎

This being said we shall focus on a simple form of Algorithm 3.7.

Algorithm 3.10 *A simple mixed preconditioner*

  1: Initialization

- r_u, r_p *given*
- $z_u = \widetilde{A}^{-1} r_u$
- $\widetilde{r}_p = B z_u - r_p$

  2: Approximation of $\widetilde{S}^{-1}$

- $z_p = M_S^{-1} \widetilde{r}_p$
- $z z_u = \widetilde{A}^{-1} B^t z_p$
- $T_p = B z z_u$
- $\beta = \dfrac{(\widetilde{r}_p, T_p)}{(T_p, T_p)}$

Fig. 3.3 Illustration of the preconditioner

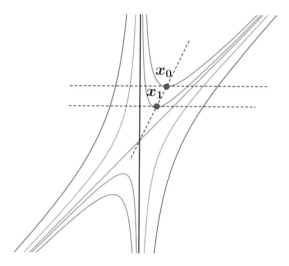

3: *Final computation*

- $z_p = \beta z_p$
- $z_u = z_u - \beta z z_u$

4: *End*

This essentially amounts to obtaining a scaling by β of $M_S^{-1} r_p$. The computation of β implies the computation of $z z_u$, which implies a resolution with the approximate solver $\tilde{A}^{-1}$. This computation would anyway be required by the last part of Algorithm 3.7 so that there is in fact no extra cost. If we have symmetry of $\tilde{A}$ and M_S, we can obtain β by one iteration of the gradient method instead of a minimum residual.

Remark 3.15 (The Preconditioner Yields a Saddle-Point) It is interesting to see in Fig. 3.3 that the preconditioner yields a saddle-point in a two-dimensional space. One can also see that $z z_u$ is A-orthogonal (conjugate) to z_u

This makes us think of the Partan method [82] where conjugation is obtained after two descent steps (see Fig. 3.4).

∎

Remark 3.16 (The Classical Method of Arrow-Hurwicz-Uzawa) We shall later use, for the sake of comparison the classical Arrow-Hurwicz-Uzawa method, described in [43], as a preconditioner.

Algorithm 3.11 *A super-simple mixed preconditioner*

1: *Input r_u, r_p, Output z_u, z_p.*
2: *Compute z_u and z_p*

- $z_u = \tilde{A}^{-1} r_u$

Fig. 3.4 Illustration of
Partan's method

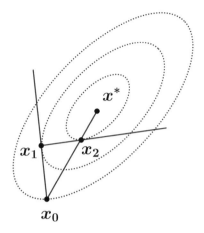

- $\tilde{r}_p = r_p - Bz_u$
- $z_p = \beta M_S^{-1}\tilde{r}_p$

The problem here is that β has to be determined by the user, while in the previous version, everything was automatic. The choice of parameters had been discussed in [43]. Moreover, as we said above, the last part of Algorithm 3.7 requires an extra resolution. ∎

Remark 3.17 (The Perturbed Problem)   We have considered in Sect. 2.1.4, a regular perturbation. We would then have to solve a problem of the form

$$\begin{pmatrix} A & B^t \\ B & -\epsilon M_Q \end{pmatrix} \begin{pmatrix} u \\ p \end{pmatrix} = \begin{pmatrix} f \\ g \end{pmatrix}$$

where M_Q is the matrix defining the metric on Q. We have noted in Remark 2.13 that S then becomes $S + \epsilon M_Q$ and that M_S is changed into $(1 + \epsilon)M_S$ if $M_Q = M_S$.

The preconditioner of Algorithm 3.10 is easily adapted to the perturbed case. If one uses $M_S = M_Q$ one should change,

- $\tilde{r}_p = Bz_u - \epsilon M_Q\, p - r_p,$
- $z_p = \dfrac{1}{1+\epsilon}M_Q^{-1}\tilde{r}_p$
- $T_p = Bz z_u - \epsilon M_Q z_p$

When the preconditioner 3.10 is employed for the modified problem of Sect. 2.1.4, one should also modify the computation of residuals in the associated CG or GCR method, taking into account for example Remark 3.9. ∎

3.2.4 Augmented Lagrangian as a Perturbed Problem

We have considered in Sect. 2.3.4 a form of the augmented Lagrangian method in which one has to solve a sequence of problems of the form (2.36), penalty problems written in mixed form.

This could be considered when a classical Augmented Lagrangian as in (2.34) would not be possible because R^{-1} is a full matrix. This could also be an alternative to the regularised lagrangian.

Using this implies a two level iteration, which can be compared to a Newton method. To fix ideas, we describe an implementation by the GCR algorithm.

- Initialise a GCR method with the residuals of the **unperturbed problem**.
- Solve for δu and δp with the penalty problem in mixed form using the preconditioner as Remark 3.17 and the GCR iteration following Remark 3.9 but without changing the initial residuals.
- Update $u + \delta u$ and $p + \delta p$.

From there, one has many variants. The internal iteration may not be too exact. The update could itself be accelerated by another GCR. When the problem is the linearised part of a Newton iteration, one could in fact use this as an inexact Newton.

Chapter 4
Numerical Results: Cases Where $Q = Q'$

We shall now illustrate the behaviour of these algorithms on some examples. In all the cases considered here, the space of multipliers Q can be identified with its dual. In fact we shall have in all cases $Q = L^2(\Omega)$.

The first example will be the mixed approximation of a Poisson problem with the simplest Raviart-Thomas element. This will allow us to consider a real discrete Augmented Lagrangian method and its impact in iterative solvers. We shall then consider incompressible problems in elasticity in both the linear and non linear cases. Finally we shall introduce an application to the Navier-Stokes equations

4.1 Mixed Laplacian Problem

The first example that we consider is the mixed formulation of a Dirichlet problem using Raviart-Thomas element. As an application, we can think of a potential fluid flow problem in porous media as in the simulation of Darcy's law in reservoir simulation. This can be seen as the paradigm of mixed methods and the simplest case of a whole family of problems. Higher order elements and applications to a mixed formulation of elasticity problems are described in [20].

4.1.1 Formulation of the Problem

Let Ω a domain of $\mathbb{R}^d$ with $d = 2$ or 3. Its boundary Γ is divide into Γ_D and Γ_N on which respectively Dirichlet and Neumann conditions will be imposed. We denote $\underline{n}$ the normal to Γ. Given $f \in L^2(\Omega)$ and $\underline{g} \in (L^2(\Gamma_N))^d$, the problem consists in

© The Author(s), under exclusive license to Springer Nature Switzerland AG 2022
J. Deteix et al., *Numerical Methods for Mixed Finite Element Problems*,
Lecture Notes in Mathematics 2318, https://doi.org/10.1007/978-3-031-12616-1_4

finding a function $p \in H^1(\Omega)$ such that

$$
\begin{cases}
-\Delta p & = f \quad \text{in } \Omega \\
p & = 0 \quad \text{on } \Gamma_D \\
(\text{grad } p) \cdot \underline{n} & = g \quad \text{on } \Gamma_N
\end{cases}
\tag{4.1}
$$

Taking $\underline{u} = \text{grad } p$, Eq. (4.1) is equivalent to

$$
\begin{cases}
\underline{u} - \text{grad } p & = 0 \quad \text{in } \Omega \\
-\text{div } \underline{u} & = f \quad \text{in } \Omega \\
p & = 0 \quad \text{on } \Gamma_D \\
\underline{u} \cdot \underline{n} & = g \quad \text{on } \Gamma_N
\end{cases}
\tag{4.2}
$$

Let

$$
V = H(\text{div}, \Omega) = \{\underline{u} \in (L^2(\Omega))^d, \ \text{div}\,\underline{u} \in L^2(\Omega)\}
$$

and

$$
Q = L^2(\Omega).
$$

The variational formulation of (4.2) is then to find $\underline{u} \in V$ and $p \in Q$ satisfying

$$
\begin{cases}
\displaystyle\int_\Omega \underline{u} \cdot \underline{v}\, dx + \int_\Omega p \,\text{div}\,\underline{v}\, dx = \int_{\Gamma_N} g \cdot \underline{v}\, ds & \forall\, \underline{v} \in V, \\
\displaystyle\int_\Omega \text{div}\,\underline{u}\, q\, dx = -\int_\Omega f\, q\, dx & \forall\, q \in Q
\end{cases}
\tag{4.3}
$$

In this context, the *inf-sup* condition [20] is verified and, the operator $a(\cdot,\cdot)$ defined on $V \times V$ by

$$
a(\underline{u}, \underline{v}) = \int_\Omega \underline{u} \cdot \underline{v}\, dx
$$

is coercive **on the kernel** of the divergence operator but not on the whole space V.
   The mixed formulation (4.3) is the optimality condition of the following *inf-sup* problem

$$
\inf_{\underline{v} \in V} \sup_{q \in Q} \frac{1}{2} \int_\Omega |\underline{v}|^2\, dx - \int_{\Gamma_N} g \cdot \underline{v}\, ds + \int_\Omega q\, \text{div}\,\underline{v}\, dx + \int_\Omega f\, q\, dx
\tag{4.4}
$$

In order to obtain a discrete form of the problem, we introduce the Raviart-Thomas elements. We suppose to have a partition $\mathcal{T}_h$ of Ω into tetrahedra. If we denote by $P_k(K)$ the polynomials of degree k on a tetrahedron K, we define,

$$RT_k(\Omega) = \{\underline{u}_h \in (P_k(K))^d + \underline{x} P_k(K) \quad \forall K \in \mathcal{T}_h\} \tag{4.5}$$

The properties of these spaces and related ones are well described in [20]. Indeed they have been built to be applied to problem (4.3). If we take

$$Q_h = \{q_h \in P_k(K) \quad \forall K \in \mathcal{T}_h\}$$

the matrix R corresponding to the scalar product on Q_h is block diagonal and we have the important property that $\operatorname{div} \underline{u}_h \in Q_h$, the **inclusion of kernels** property which ensures that coercivity on the kernel is also valid for the discrete problem. The discrete operator div_h corresponding to the definition of (2.16) is indeed the restriction of div to V_h.

4.1.2 Discrete Problem and Classic Numerical Methods

The discrete problem is clearly of the form (2.22) and indefinite. In [20, p. 427], this was considered 'a considerable source of trouble'. Let us consider things in some detail.

- The matrix A is built from a scalar product in $L^2(\Omega)$.
- The operator B is the standard divergence
- B^t is a kind of finite volume gradient.
- The Schur complement $BA^{-1}B^t$ is a (strange) form of the Laplace operator acting on piecewise constant.

It was shown in [12] how using a diagonalised matrix for the matrix A, one indeed obtains a finite volume method. In the two-dimensional case, it is known (see [68]) that the solution can be obtained from the non-conforming discretisation of the Laplacian.

Another general approach (see [6] for example) is to impose the interface continuity of the normal components in V_h by Lagrange multipliers to generate a positive definite form.

We shall rather stick to the indefinite formulation an show how the methods that we developed in Chap. 3 can be applied.

The Augmented Lagrangian Formulation

The first point is that we have a lack of coercivity on the whole space. The result of Proposition 2.2 does not hold and the convergence of Uzawa's method, for example would not be independent of h. To obtain the coercivity on the whole of V we shall consider an augmented Lagrangian method, which also means to use

$$a(\underline{u}, \underline{v}) + \alpha(\operatorname{div}\underline{u}, \operatorname{div}\underline{v}).$$

Solving the equation with the augmented Lagrangian method gives us the *inf-sup* problem

$$\inf_{\underline{v}} \sup_{q} \frac{1}{2} \int_{\Omega} |\underline{v}|^2 \, dx + \frac{\alpha}{2} \int_{\Omega} |\operatorname{div}\underline{v} + f|^2 \, dx$$
$$- \int_{\Gamma_N} \underline{g} \cdot \underline{v} \, ds + \int_{\Omega} q \operatorname{div}\underline{v} \, dx + \int_{\Omega} f q \, dx \tag{4.6}$$

where $\alpha > 0$ is a scalar representing the parameter of regularisation. It's well known that the equation (4.4) is equivalent to (4.6) for which optimality conditions are

$$\begin{cases} \displaystyle\int_{\Omega} \underline{u} \cdot \underline{v} \, dx + \alpha \int_{\Omega} \operatorname{div}\underline{u} \operatorname{div}\underline{v} \, dx + \int_{\Omega} p \operatorname{div}\underline{v} \, dx \\ \qquad = \displaystyle\int_{\Gamma_N} \underline{g} \cdot \underline{v} \, ds - \alpha \int_{\Omega} f \operatorname{div}\underline{v} \, dx \qquad \forall \underline{v} \in V, \\ \displaystyle\int_{\Omega} \operatorname{div}\underline{u} \, q \, dx = -\int_{\Omega} f q \, dx \qquad\qquad\qquad \forall q \in Q. \end{cases} \tag{4.7}$$

Here, for all α greater than zero, the coercivity is satisfied on V since

$$a(\underline{v}, \underline{v}) + \alpha \int_{\Omega} \operatorname{div}\underline{u} \operatorname{div}\underline{v} \, dx \geq \min(1, \alpha^2)\|\underline{v}\|_V^2, \qquad \forall \underline{v} \in V,$$

with

$$\|\underline{v}\|_V^2 = \left(\int_{\Omega} |\underline{u}|^2 \, dx + \int_{\Omega} |\operatorname{div}\underline{v}|^2 \, dx \right)$$

A consequence is that the two forms of the augmented Lagrangian (2.30) and (2.32) are the same and that using them will not change the solution of the problem.

Obviously (2.32) corresponds to the linear system of equations associated to the discrete version of (4.7). As we said earlier, for the Raviart-Thomas elements $M_S = R$ is a block diagonal matrix associated to the L^2 scalar product on Q_h and we are in the perfect situation (Remark 2.11) for the augmented Lagrangian.

$$\begin{pmatrix} A + \alpha B^t R^{-1} B & B^t \\ B & 0 \end{pmatrix} \begin{pmatrix} u \\ p \end{pmatrix} = \begin{pmatrix} g - \alpha B^t M_D^{-1} f \\ f \end{pmatrix}.$$

4.1.3 A Numerical Example

For our numerical tests, we take $k = 0$ in (4.5). Q_h is thus a space of piecewise constants. For the numerical example, we consider the domain Ω as a cube of unit edges. We take Γ_D empty so $\Gamma_N = \partial\Omega$. The functions f and g are given by:

$$f(x, y, z) = -3\pi^2 \sin(\pi x) \sin(\pi y) sin(\pi z)$$

$$g(x, y, z) = 0$$

We use a tolerance of 10^{-12} for the convergence of the algorithms.

The interest of this example is to see the effect of the augmented Lagrangian formulation and the role of the parameter α. We would like to emphasise again that we are in an ideal case where the augmented Lagrangian does not change the solution.

- Small values of α act as regularisation parameter. They provide coercivity on the whole of V and not only on the kernel of the divergence operator.
- Large values of α yield a penalty method which is then corrected by an iteration. This is the classical augmented Lagrangian method.

To evaluate the effect of the augmented Lagrangian we first present results in which we use a relatively crude mesh with 50 688 tetrahedra and 24 576 faces.

We first consider the case of a direct solver for the primal problem in u. This is therefore the Uzawa algorithm of Sect. 3.2.2. The number of iterations in p depends on the condition number of the dual problem (2.23) which improves when α becomes large. This is observed in our experiences. Table 4.1 illustrate that a larger α yields a better convergence as should be expected. In fact we see that the largest improvement is to have coercivity on the whole space (taking $\alpha > 0$) rather then having coercivity only on the kernel ($\alpha = 0$). Notice that increasing values of α does not improve things as the condition number of the problem in u becomes bad.

We can compare this with an incomplete resolution in u. This is done by using a simple iterative method: a fixed number of the conjugate gradient method with an SSOR preconditioner, which we denote CG(n). The results presented in Table 4.2 were obtained for $n = 5$ and $n = 10$.

We observe once again a large difference between $\alpha = 0$ and $\alpha > 0$. We also see that there is an optimal α and that for too large values the condition number

Table 4.1 Laplace problem with coarse mesh and complete (LU) resolution for the primal problem: global number of iterations and computing time in seconds according to the value of α

α	# iterations	CPU time (s)
0	133	18.57
10^{-1}	7	3.2995
10^2	5	2.8836
10^5	3	2.5867
10^7	3	2.5305

Table 4.2 Laplace problem with coarse mesh with CG(5)-SOR and CG(10)-SOR for the primal solver: number global of iterations (# it.) and computing time in seconds according to the value of α

α	CG(5)-SOR		CG(10)-SOR	
	# it.	CPU time(s)	# it.	CPU time(s)
0	133	13.6105	133	14.8259
1.0×10^{-4}	80	7.0878	79	8.8246
5.0×10^{-4}	45	4.2140	42	5.1697
1.0×10^{-3}	38	3.7885	31	4.2228
2.5×10^{-3}	37	3.9470	25	3.7092
5.0×10^{-3}	37	3.7123	24	3.6260
1.0×10^{-2}	47	4.4617	22	3.4336

Table 4.3 Laplace problem with fine mesh: total number of iterations and computing time in seconds according to the solver of primal problem

	LU $(\alpha = 10^5)$	CG(100) $(\alpha = 2.5 \times 10^{-3})$	CG(50) $(\alpha = 10^{-2})$
# it.	3	15	25
CPU (s)	808.5	586.7	569.1

in $\underline{u}$ worsens the performance. If we improve the resolution, we have indeed better results.

With a better solver, the optimal value of α increases and the number of iterations in p decreases. However, the computing time does not as each iteration becomes more expensive.

Since the goal is to solve big problems, we took a rather fine mesh (with 3,170,304 faces and 1,572,864 tetrahedra) to prove the interest of the algorithm with an incomplete resolution. In this context, the direct resolution is very expensive in time as can be seen in Table 4.3 whereas in iterative resolution, the computation time is less.

We can see that an iterative method for $\underline{u}$ can be better than a direct solver. This result could surely be much improved by a more clever choice of this iterative method.

Remark 4.1 (This Is not Optimal) We would like to emphasise that the choice of the solver in $\underline{u}$ is not optimal and that one could improve the results. Our main point is that the indefinite form of mixed methods is not such a 'source of trouble'!   ■

4.2   Application to Incompressible Elasticity

We consider a second class of examples, arising from the discretisation of incompressible elasticity problems. We consider two cases : a standard linear elasticity model and a Mooney-Rivlin model [21, 53, 84]. This relatively simple case will allow us to compare the preconditioners introduced and to discuss strategies to improve coercivity.

4.2.1   *Nearly Incompressible Linear Elasticity*

Given a body Ω, we want to determine its displacement $\underline{u}$ under external forces. Denoting Γ the boundary of Ω, we define Γ_D the part of Γ on which Dirichlet (displacements) conditions are imposed and Γ_N the part where we have Neumann (forces) conditions. To avoid unnecessary developments, we restrict ourselves to the case of null conditions on Γ_D. We thus define,

$$V = \{\underline{v} \,|\, \underline{v} \in (H^1(\Omega))^d, \ \underline{v} = 0 \text{ on } \Gamma_D\}$$

For $\underline{v} \in V$, we define the linearised strain tensor

$$\underline{\underline{\varepsilon}}_{ij}(\underline{v}) = \frac{1}{2}(\partial_i u_j + \partial_j u_i) \tag{4.8}$$

and its deviator

$$\underline{\underline{\varepsilon}}^D = \underline{\underline{\varepsilon}} - \frac{1}{3} tr(\underline{\underline{\varepsilon}})\mathbf{I}$$

One then has,

$$|\underline{\underline{\varepsilon}}^D(\underline{v})|^2 = |\underline{\underline{\varepsilon}}(\underline{v})|^2 - \frac{1}{3} tr(\underline{\underline{\varepsilon}}(\underline{v}))^2 \tag{4.9}$$

To define our problem, we have to define some parameters. Elasticity problems are usually described by the Young Modulus E and the Poisson ratio v. We shall rather employ the Lamé coefficients μ, λ

$$\mu = \frac{E}{2(1+v)}, \quad \lambda = \frac{Ev}{(1+v)(1-2v)} \tag{4.10}$$

and the bulk modulus k

$$k = \frac{E}{3(1-2v)} = \lambda + \frac{2\mu}{3} \tag{4.11}$$

We then consider the problem

$$\inf_{\underline{v} \in V} \mu \int_\Omega |\underline{\underline{\varepsilon}}^D(\underline{v})|^2 \, dx + \frac{k}{2} \int_\Omega |\operatorname{div} \underline{v} - g|^2 \, dx - \int_\Omega \underline{f} \cdot \underline{v} \, dx \tag{4.12}$$

or equivalently,

$$\inf_{\underline{v} \in V} \mu \int_\Omega |\underline{\underline{\varepsilon}}(\underline{v})|^2 \, dx + \frac{\lambda}{2} \int_\Omega |\operatorname{div} \underline{v} - g|^2 \, dx - \int_\Omega \underline{f} \cdot \underline{v} \, dx. \tag{4.13}$$

Problem (4.13), for example, leads to the variational form,

$$2\mu \int_\Omega \underline{\underline{\varepsilon}}(\underline{u}) : \underline{\underline{\varepsilon}}(\underline{v}) \, dx + \lambda \int_\Omega (\text{div}\,\underline{u} - g) \,\text{div}\,\underline{v}\, dx = \int_\Omega \underline{f} \cdot \underline{v}\, ds \quad \forall\, \underline{v} \in V.$$
$$(4.14)$$

It it is well known that a brute force use of (4.14) or (4.13) could lead to bad results for large values of λ (or as the Poisson ratio ν nearing its maximal value of $1/2$). In extreme cases, one gets a locking phenomenon that is an identically zero solution.

The standard way to circumvent this locking phenomenon is to switch to a mixed formulation with a suitable choice of elements. Essentially, we introduce the variable

$$p = \lambda(\text{div}\,\underline{v} - g)$$

and we consider the saddle-point problem,

$$\inf_{\underline{v}} \sup_{q} \mu \int_\Omega |\underline{\underline{\varepsilon}}(\underline{v})|^2 \, dx - \frac{1}{2\lambda} \int_\Omega |q|^2\, dx$$
$$+ \int_\Omega q\,(\text{div}\,\underline{v} - g)dx - \int_\Omega \underline{f} \cdot \underline{v}\, ds$$

for which optimality conditions are and denoting $(\underline{u}, p) \in V \times Q$ the saddle point,

$$\begin{cases} 2\mu \int_\Omega \underline{\underline{\varepsilon}}(\underline{u}) : \underline{\underline{\varepsilon}}(\underline{v}) \, dx + \int_\Omega p\,\text{div}\,\underline{v}\, dx = \int_\Omega \underline{f} \cdot \underline{v}\, ds & \forall\, \underline{v} \in V, \\[2mm] \int_\Omega \text{div}\,\underline{u}\, q\, dx - \dfrac{1}{\lambda} \int_\Omega pq\, dx = \int_\Omega g\,qdx & \forall\, q \in Q \end{cases}$$
$$(4.15)$$

In the limiting case of λ becoming infinite, the case that we want to consider, the second equation of (4.15) becomes

$$\int_\Omega \text{div}\,\underline{u}\, q\, dx = \int_\Omega g\,qdx \quad \forall\, q \in Q$$

Problem (4.15) is well posed. Indeed we have,

- The bilinear form $a(\underline{u}, \underline{v}) = \int_\Omega \underline{\underline{\varepsilon}}(\underline{u}) : \underline{\underline{\varepsilon}}(\underline{v}) \, dx$ is coercive on V. This is Korn's inequality : there exist a constant α such that

$$a(\underline{v}, \underline{v}) = \mu(|\underline{\underline{\varepsilon}}(\underline{v})|^2) \geq \alpha \|\underline{v}\|_V^2 \quad \forall\, \underline{v} \in V$$

- We have an Inf-sup condition [20]: there exists a constant β such that

$$\sup_{\underline{v} \in V} \frac{b(\underline{v}, q)}{\|\underline{v}\|} \geq \beta \|q\| \quad \forall q \in Q.$$

Remark 4.2 If we use formulation (4.12) instead of (4.13) as the starting point, the bilinear form

$$a^D(\underline{u}, \underline{v}) = \int_\Omega \underline{\underline{\varepsilon}}^D(\underline{u}) : \underline{\underline{\varepsilon}}^D(\underline{v}) \, dx$$

is coercive on the kernel of B. Indeed, by Korn's inequality and (4.9) : there exist a constant α such that

$$\mu |\underline{\underline{\varepsilon}}^D(\underline{v})|^2 = \mu \left(|\underline{\underline{\varepsilon}}(\underline{v})|^2 - \frac{1}{3} tr(\underline{\underline{\varepsilon}}(\underline{v}))^2 \right) \geq \alpha \|\underline{v}\|^2 - \frac{\mu}{3} tr(\underline{\underline{\varepsilon}}(\underline{v}))^2) \quad \forall \underline{v} \in V$$

This makes it possible for the matrix A^D defined by the bilinear form $a^D(\cdot, \cdot)$ to be singular, a situation which is not be acceptable in our algorithms. ∎

Remark 4.3 In the limiting case of infinite λ, that is of incompressible materials, problems (4.12) and (4.13) are equivalent as we have $div \ \underline{u} = tr(\underline{\underline{\varepsilon}}(\underline{u})) = 0$, the solution is unchanged and for (4.13) we have coercvity on the whole of V and not only on the kernel of B. However, for the **discretised problem**, the solution is modified as in general one does not have $div \ \underline{u}_h = tr(\underline{\underline{\varepsilon}}(\underline{u}_h)) = 0$. ∎

In our numerical tests, we shall use the Augmented Lagrangian but as in (2.30) that is a 'regularised Lagrangian'

$$
\begin{cases}
2\mu \int_\Omega \underline{\underline{\varepsilon}}(\underline{u}) : \underline{\underline{\varepsilon}}(\underline{v}) \, dx + \hat{\lambda} \int_\Omega (div \, \underline{u} - g) \ div \, \underline{v} \, dx \\
\qquad\qquad + \int_\Omega p \ div \, \underline{v} \, dx = \int_\Omega \underline{f} \cdot \underline{v} \, ds = 0 \qquad \forall \underline{v} \in V, \\
\int_\Omega div \, \underline{u} \, q \, dx = \int_\Omega g \, q dx \qquad\qquad\qquad \forall q \in Q
\end{cases}
$$

$$(4.16)$$

Where we define $\hat{\lambda}$ using an artificial Poisson ratio $\hat{\nu}$ as in (4.10) which should in no way be close to $1/2$ unless the choice of elements allows really divergence-free solution. The parameter $\hat{\lambda}$ is arbitrary and is chosen to obtain a good convergence without downgrading the condition number of the system. This will be discussed in Sect. 4.2.5.

Remark 4.4 (The Stokes Problem) It is clear that all we say about this linear elasticity problem is valid for the equivalent Stokes problem for creeping flow

problems. The numerical solution of the Stokes problem has been the object of a
huge number of articles and books.   ∎

4.2.2   Neo-Hookean and Mooney-Rivlin Materials

We now consider non linear elasticity models described in a Lagrangian formulation. We thus have a reference domain denoted Ω and a deformed domain ω while
their respective boundaries are denoted as Γ and γ. We denote X and x respectively
their coordinates. We write the deformation of a body as

$$x = \underline{u}(X)$$

The deformation $\mathbf{F}$ of this transformation is given by $\mathbf{F} = \dfrac{\partial x}{\partial X} = \mathbf{I} + \nabla_X \underline{u}$ and its
determinant $det\mathbf{F}$ is denoted J. Note that ∇_X stands for the gradient with respect
to the variable X. The Cauchy-Green tensor is then defined as $\mathbf{C} = \mathbf{F}^T \mathbf{F}$ and its
principal invariants I_1, I_2 and I_3 are given by :

$$I_1 = \mathbf{C} : \mathbf{I}, \quad I_2 = \frac{1}{2}(I_1^2 - \mathbf{C} : \mathbf{C}), \quad I_3 = \det(\mathbf{C}) = J^2.$$

As in the case of linear elasticity, the boundary Γ is composed of (at least) two
parts: Γ_D where a Dirichlet condition is given and Γ_N where a Neumann (pressure)
condition g is imposed.

- **The Neo-Hookean model**

    Although there are many formulations of Neo-Hookean models for compress-
ible materials, they share a unique *elastic potential energy* function or *strain
energy* function W as named in [73]. Following [91], we define a particular
Neo-Hookean material where the potential energy function W is given by

$$W = \frac{\mu}{2}(I_1 - 3) + \frac{\lambda}{4}(J^2 - 1) - (\frac{\lambda}{2} + \mu) \ln J$$

$$= \frac{\mu}{2}\left(I_3^{-\frac{1}{3}} I_1 - 3\right) + \frac{1}{2}\kappa(J - 1)^2$$

    where the material is characterized by μ and λ the Lamé coefficients as in the
linear model and κ the bulk modulus.

- **The Mooney-Rivlin model**

    For a Mooney-Rivlin hyperelastic material, the energy functional is given by

$$W = \mu_{10}\left(I_3^{-\frac{1}{3}} I_1 - 3\right) + \mu_{01}\left(I_3^{-\frac{2}{3}} I_2 - 3\right) + \frac{1}{2}\kappa(J - 1)^2$$

where κ, the bulk modulus, μ_{10} and μ_{01} are parameters characterizing the material.

The elasticity problem consists in minimizing the potential energy W under appropriate boundary conditions. The weak formulation on the reference configuration Ω can be written as a nonlinear problem,

$$\int_\Omega (\mathbf{F} \cdot \mathbf{S}) : \nabla_X \underline{v} \, dX = \int_\Omega \underline{f} \cdot \underline{v} \, dX + \int_{\Gamma_N} \underline{g} \cdot \underline{v} \, d\gamma \qquad (4.17)$$

for any $\underline{v}$ in a proper functional space and where $\mathbf{S} = 2\partial W / \partial \mathbf{C}$ is the second Piola-Kirchoff stress tensor. More details on the formulation can be found in [36].

Mixed Formulation for Mooney-Rivlin Materials

We are interested in the simulation of incompressible materials, more particularly rubberlike materials. As in the case of linear materials, it is well known that displacement-only formulations are inadequate. As the material becomes more and more incompressible (corresponding to an increasing bulk modulus), the conditioning of the linearized system grows. And the usual locking phenomena may occur when the bulk modulus κ becomes large. As a consequence, direct solvers and iterative solvers are ineffective (even with preconditionning).

It is then convenient to split the second Piola-Kirchoff tensor $\mathbf{S}$ into a volumic part and an isochoric part and to use a mixed formulation in which p is explicitly an unknown. We thus write

$$\mathbf{S} = \mathbf{S}' - pJ\mathbf{C}^{-1}.$$

Here p is the pressure which is defined by $p = -\kappa(J - 1)$. For a Mooney-Rivlin hyperelastic material (of which a Neo-Hookean material is a special case),

$$\mathbf{S}' = 2\mu_{10} I_3^{-1/3} \left(\mathbf{I} - \frac{1}{3} I_1 \mathbf{C}^{-1} \right) + 2\mu_{01} I_3^{-2/3} \left(I_1 \mathbf{I} - \mathbf{C} - \frac{2}{3} I_2 \mathbf{C}^{-1} \right)$$

The mixed formulation can now be written as:

$$\begin{cases} \int_\Omega (\mathbf{F} \cdot \mathbf{S}') : \nabla_X \underline{v} \, dX - \int_\Omega pJ\mathbf{F}^{-T} : \nabla_X \underline{v} \, dX \\ \qquad\qquad = \int_{\Gamma_N} \underline{g} \cdot \underline{v} \, dS + \int_\Omega \underline{f} \cdot \underline{v} \, dX \qquad \forall \underline{v} \in V \qquad (4.18) \\ \int_\Omega (J - 1)q \, dX + \frac{1}{\kappa} \int_\Omega pq \, dX = 0 \qquad\qquad \forall q \in Q \end{cases}$$

We are interested in the really incompressible case when the bulk modulus becomes infinite. As we have seen in the linear case, this may lead to an ill conditioned or even singular matrix in $\underline{u}$. To get good results, we shall again introduce a stabilisation parameter $\hat{K}$. We then define $\hat{S}$ using this artificial small bulk modulus. and solve

$$
\begin{cases}
\displaystyle\int_\Omega (\mathbf{F} \cdot \hat{\mathbf{S}}) : \nabla_X \underline{v}\, dX - \int_\Omega pJ\mathbf{F}^{-T} : \nabla_X \underline{v}\, dX \\[2ex]
\qquad\qquad \displaystyle\int_{\Gamma_N} \underline{g} \cdot \underline{v}\, dS + \int_\Omega \underline{f} \cdot \underline{v}\, dX \qquad \forall \underline{v} \in V \qquad (4.19) \\[2ex]
\displaystyle\int_\Omega (J - 1)q\, dX = 0 \qquad\qquad\qquad\qquad \forall q \in Q
\end{cases}
$$

This is a non linear formulation, its resolution will be based on a Newton-like method and will imply a linearisation of (4.18). To lighten the presentation (we refer to [42] for the derivation of the linearised problem), let us rewrite the non linear problem (4.18) as

$$
\begin{cases}
R_1((\underline{u}, p), \underline{v}) = 0, \\
R_2((\underline{u}, p), q) = 0.
\end{cases}
$$

At each Newton's iterations, we consider a linearized version of this system with respect to $\underline{u}$ and p. Linearizing around $(\underline{u}_n, p_n)$, we get the followings bilinears operators

$$
\begin{aligned}
a_n(\delta\underline{u}, \underline{v}) &= \frac{\partial R_1((\underline{u}_n, p_n), \underline{v})}{\partial \underline{u}} \cdot \delta\underline{u} \\[2ex]
&= \int_\Omega \mathbf{S}'(\underline{u}_n) : \left(\nabla_X^{\mathrm{T}}(\delta\underline{u}) \cdot \nabla_X \underline{v}\right) dX \\[2ex]
&\quad + \int_\Omega \left(C(\underline{u}_n) : \left(\mathbf{F}^{\mathrm{T}}(\underline{u}_n) \cdot \nabla_X(\delta\underline{u})\right)\right) : \left(\mathbf{F}^{\mathrm{T}}(\underline{u}_n) \cdot \nabla_X \underline{v}\right) dX,
\end{aligned}
$$

$$
\begin{aligned}
b_n(\underline{v}, \delta p) &= \frac{\partial R_1((\underline{u}_n, p_n), \underline{v})}{\partial p} \cdot \delta p \\[2ex]
&= -\int_\Omega J\delta p\left(\mathbf{F}^{-\mathrm{T}}(\underline{u}_n) : \nabla_X \underline{v}\right) dX
\end{aligned}
$$

$$
\begin{aligned}
c_n(\delta p, q) &= \frac{\partial R_2((\underline{u}_n, p_n), q)}{\partial p} \cdot \delta p \\[2ex]
&= -\int_\Omega \frac{1}{k}(\delta p)q\, dX.
\end{aligned}
$$

The linearised variational formulation is, knowing $(\underline{u}_n, p_n)$, the previous solution, to find $(\delta\underline{u}, \delta p)$ such that

$$
\begin{cases}
a_n(\delta\underline{u}, \underline{v}) + b_n(\underline{v}, \delta p) = -R_1\Big((\underline{u}_n, p_n), \underline{v}\Big), \ \forall \underline{v} \in V, \\
b_n(\delta\underline{u}, q) - c_n(\delta p, q) = -R_2\Big((\underline{u}_n, p_n), q\Big), \ \forall q \in Q.
\end{cases}
\tag{4.20}
$$

Remark 4.5 An important point is that the linearised system (4.20) depends on some initial value of the displacement. In general, we do not have the equivalent of Korn's inequality and the matrix can in fact be singular if one has chosen as initial guess a bifurcation point.   ∎

4.2.3   Numerical Results for the Linear Elasticity Problem

To illustrate the behaviour of our algorithms, we first consider the simple case of linear elasticity on a three-dimensional problem. The results will also be applicable to the Stokes problem.

To obtain an approximation of (4.16), we have to choose a finite element space $V_h \subset V$ and a space $Q_h \subset Q$. In our numerical experiments, we consider a three-dimensional problem. We thus had to make a choice of a suitable finite element approximation. The catalogue of possibilities has been well studied [20] and we made a choice which seemed appropriate with respect to the standard engineering applications. We employ tetrahedral elements, a choice motivated by our eventual interest in mesh adaptation and we want to respect the inf-sup condition without having to use elements of too high degree. The popular Taylor-Hood element was retained:

- A piecewise quadratic approximation for the displacement $\underline{u}$ and a piecewise linear approximation of the pressure p.

Remark 4.6 This choice of element is good but there is a restriction for the construction of element at the boundary: no element should have all its vertices on the boundary. This might happen on an edge if no special care is taken when the mesh is built. A bubble can be added at the displacement to avoid this restriction, but to the price of more degrees of freedom.   ∎

Remark 4.7 (Regularised Lagrangian)   We therefore have a **continuous approximation** for pressure and this has some consequences on the choice of solvers, in particular if one would like to use the augmented Lagrangian methods. It is almost impossible to employ (2.32) as for all reasonable choices of M_S, M_S^{-1} is a full matrix. However, as we shall see later the regularised formulation (2.30) term may improve convergence.

In order to use the real augmented Lagrangian (2.32) , one would need a **discontinuous pressure** element, which would make $M_S = R$ block diagonal. For three dimensional problems such elements are of high polynomial degree [20] or induce a loss in the order of convergence.   ■

We therefore consider the discrete regularised Lagrangian corresponding to (4.16)

$$
\begin{cases}
2\mu \int_\Omega \underline{\underline{\varepsilon}}(\underline{u}_h) : \underline{\underline{\varepsilon}}(\underline{v}_h)\, dx + \hat{\lambda} \int_\Omega (\operatorname{div} \underline{u}_h - g)\operatorname{div} \underline{v}_h\, dx \\
\qquad\qquad + \int_\Omega p_h\, \operatorname{div} \underline{v}_h\, dx = \int_\Omega \underline{f} \cdot \underline{v}_h\, ds = 0 \quad \forall \underline{v}_h \in V_h, \qquad (4.21) \\
\int_\Omega \operatorname{div} \underline{u}_h\, q_h\, dx = \int_\Omega g\, q_h dx \qquad\qquad\qquad \forall q_h \in Q_h
\end{cases}
$$

We shall later discuss in detail the choice of $\hat{\lambda}$.

4.2.4   The Mixed-GMP-GCR Method

In the following, we employ Algorithm 3.5 (Mixed-P-GCR) using the general mixed precondiioner GMP of Algorithm 3.7. The approximation of S is the one iteration version Algorithm 3.10. We shall call this combination Mixed-GMP-GCR. To complete the description, we must define a solver in $\underline{u}$ for which we consider different possibilities.

Approximate Solver in $\underline{u}$

Our mixed solver relies on an approximate solver for the problem in $\underline{u}$. We shall explore various possibilities for this choice, corresponding to different choice for $\tilde{A}^{-1}$ in (3.8).

- The direct solver is denoted LU.
- The conjugate gradient method CG, GMRES or GCR methods can also be used.
- We also consider the HP method of [37] already discussed in Sect. 3.1.4. In this method, the quadratic approximation P_2 is split into a linear P_1 part defined on the vertices and a complement P_2 part defined on edges and the matrix A is split into four submatrices

$$
A = \begin{pmatrix} A_{11} & A_{12} \\ A_{21} & A_{22} \end{pmatrix}
$$

From this splitting, we obtain a two-level algorithm in which one solves in sequence the P_1 part and the P_2 part. From [89], we know that the P_2 part is well conditioned so that a simple iteration is suitable,

- For the P_1 part, we use an Algebraic Multigrid Method (AMG) or a direct LU factorisation.
- For the P_2 part, we use a SSOR iteration.

HP-AMG will denote the use of the HP solver with AMG; HP-LU the use of the HP solver with LU. We then use these solvers, as a preconditionner in the following ways :

- as a preconditioner on its own, limited to one iteration, denoted as PREONLY,
- as a preconditioner for a **GCR** method or a **GMRES** method (see below).

With n the (limited) number of iterations of the GCR or GMRES, we shall denote GCR(n)-HP-AMG, GCR(n)-HP-LU, GMRES(n)-HP-AMG or GMRES(n)-HP-LU the approximate solver for A using one iteration of the HP-AMG or HP-LU solver as a preconditioner.

These approximate solver of A are used in a general mixed preconditioner (GMP, here we use Algorithm 3.10) and the global system is solved using a Mixed-GMP-GCR solver (Algorithm 3.5).

Remark 4.8 For the approximation M_S of the Schur's complement in Algorithm 3.10, we take the matrix associated with the L^2 scalar product which we denote M_0. Computing M_0^{-1} can be solved by a direct or iterative solver. For the numerical tests, we use the LU factorisation (which can be done once for all). ■

Remark 4.9 (Variable Coefficients)   Related problems for non Newtonian flows lead to variable coefficients. In [48] one considers the choice of the approximation M_S (Sect. 2.2.3) to the Schur complement, which also defines a scalar product on Q_h. They show that if the bilinear form

$$2\mu \int_\Omega \underline{\underline{\varepsilon}}(\underline{u}) : \underline{\underline{\varepsilon}}(\underline{v}) \, dx$$

is changed into

$$2 \int_\Omega \mu(x)\underline{\underline{\varepsilon}}(\underline{u}) : \underline{\underline{\varepsilon}}(\underline{v}) \, dx \tag{4.22}$$

one should take for M_S

$$M_S = \int \frac{1}{\mu(x)} \, p_h q_h dx.$$

From Sect. 2.3 one should change the regularised form (4.16) into

$$2 \int_\Omega \mu(x)\underline{\underline{\varepsilon}}(\underline{u}) : \underline{\underline{\varepsilon}}(\underline{v}) \, dx + \hat{\lambda} \int_\Omega \mu(x)(\mathrm{div}\,\underline{u} - g) \, \mathrm{div}\,\underline{v} \, dx$$

$$+ \int_\Omega p \, \mathrm{div}\,\underline{v} \, dx = \int_\Omega \underline{f} \cdot \underline{v} \, ds = 0 \quad \forall \underline{v} \in V$$

■

4.2.5 The Test Case

We consider a simple academic example : a cube $[0, 1]^3$ clamped at the bottom (the plane $z = 0$) is submitted to a vertical displacement imposed on the top (plane $z = 1$) (see Fig. 4.1).

We consider a linear **incompressible** material with a Young's modulus $E = 10^2 \, MPa$ and, depending on the numerical experiment, an artificial Poisson coefficient $\hat{\nu}$ varying between 0.0 and 0.4. Four meshes of sizes $h = 0.5, 0.25, 0.125, 0.0625$ (respectively 2 187, 14 739, 107 811 and 823 875 degrees of freedom) will be considered. Exceptionally, for Fig. 4.3, a very coarse mesh ($h = 0.5$) will also be used. Although this can be seen as a simple problem, it must be noted that it is a true three-dimensional case.

Remark 4.10 We shall present some examples illustrating the use of our mixed solvers. In these experiments, we have imposed a rather strict tolerance of 10^{-10} on the l^2 norm of the residual in p.   ■

We have introduced in (4.21) a 'regularised formulation' parametrised by $\hat{\lambda}$ which we may associate with an artificial Poisson ratio $\hat{\nu}$. We emphasise that this is not an Augmented Lagrangian: the penalty term is introduced for the **continuous** divergence-free condition and not for the discrete one so that, for the discretisation that we employ, the penalty parameter must be small in order not to perturbate the

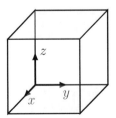

$$\begin{cases} u(x, y, 1) := [0, 0, -2] \\ u(x, y, 0) := [0, 0, 0] \end{cases}$$

Fig. 4.1 Geometry and boundary conditions

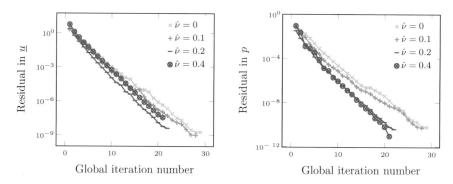

Fig. 4.2 Linear elasticity problem with GCR(3)-HP-AMG as primal solver: convergence in l^2-norm of the primal and dual residuals according to the artificial Poisson ratio $\hat{\nu}$

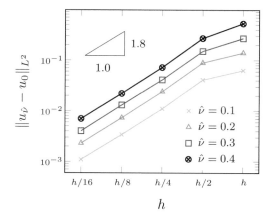

Fig. 4.3 Linear elasticity problem with GCR(3)-HP-AMG as primal solver: $\|u_{\hat{\nu}} - u_0\|_{L^2}$ with respect h according to the value of $\hat{\nu}$

solution. The global system is solved by Mixed-GMP-GCR as described above . The problem in $\underline{u}$ by three iterations of GCR with a preconditioning by HP-AMG.

Figure 4.2 shows that increasing $\hat{\nu}$ yields to a better convergence in p but has an adverse effect on the convergence in $\underline{u}$ when $\hat{\nu}$ is larger than 0.3. If we take the solution for $\hat{\nu} = 0$ (the non-regularised problem) as a reference, we may quantify the effect of the regularising parameter $\hat{\nu}$.

Figure 4.3 illustrate this effect using the difference in l^2-norm of $\underline{u}$ for various values of $\hat{\nu}$ with respect to the mesh size h (from 0.2 to 0.0125). One sees in Fig. 4.3 that the regularised formulation changes the solution but still yields convergence to the same solution when h goes to zero. It must be noted that the convergence is not quadratic as expected for a smooth solution. In fact the solution of our test problem is not regular [32] and the slope 1.8 corresponds to this lack of regularity.

Fig. 4.4 Linear elasticity problem: convergence in l^2-norm of the dual residuals according to the penalty value ϵ

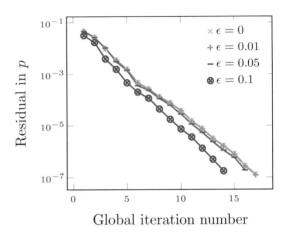

Remark 4.11 (Iterated Exact Augmented Lagrangian) In Sect. 2.3.4 we presented a way of using an exact augmented Lagrangian method, instead of a regularised one, avoiding the matrix $B^t M_S^{-1} B$ which cannot be handled as M_S^{-1} is a full matrix with our choice of elements. This implies solving the problem

$$\begin{pmatrix} A & B^t \\ B & -\epsilon M_S \end{pmatrix} \begin{pmatrix} \delta u \\ \delta p \end{pmatrix} = \begin{pmatrix} r_u \\ r_p \end{pmatrix}. \tag{4.23}$$

We then consider the same test case and following Sect. 3.17, we solve by the same algorithm which we used for the regularised problem.

As we had discussed in Sect. 2.3.4, one sees in Fig. 4.4 that increasing ϵ accelerates the convergence in p. In this test the acceleration does not justify the extra iteration in δu, δp. we conclude that, at least with the solver employed, this method would be useful only if the acceleration of the convergence of p is very important.   ∎

Number of Iterations and Mesh Size

When the solver in $\underline{u}$ is LU, the algorithm becomes the standard Uzawa method (Sect. 3.2.2). For the problem that we consider, the number of iterations in p is then independent of the mesh size as is the condition number of the dual problem (see [90]). It is interesting that this property holds even with our iterative solver in $\underline{u}$ as can be seen in Table 4.4

Table 4.4 Linear elasticity problem with GCR(3)-HP-AMG as primal solver: global number of iterations with $v = 0.4$ according to the size of the mesh

Number of degrees of freedom	2187	14,739	107,811	823,875
Number of iterations	21	22	22	21

Table 4.5 Elasticity problem: Algorithm 3.7 method. Performance with optimal (automated) parameter β

Value of n (CG(n))	1	2	3	4	5
# iterations	662	188	134	130	117
CPU time (s)	128	53	44	52	61

Comparison of the Preconditioners of Sect. 3.2

In the previous computations, we employed a combination of methods which were indeed quite efficient. One could wonder if this choice is the best. Obviously we will not definitively answer to this question, but we can certainly illustrate the merit of specific mixed methods built from Algorithm 3.5.

We first try to establish whether

- the use of Algorithm 3.7 as a solver in itself can be considered and if the GCR really accelerates.
- the cost of computing the optimal parameter β in Algorithm 3.10 is worth the effort of a second resolution in $\underline{u}$.

We thus still consider Algorithm 3.7 but we use three variants.

- Algorithm 3.7 as an iterative solver with Algorithm 3.10 for $\widetilde{S}$,
- Algorithm 3.7 as an iterative solver with Algorithm 3.11 for $\widetilde{S}$,
- Algorithm 3.7 using Algorithm 3.10 as a preconditioner for a GCR method.

We want to compare these three variants with respect to numerical behaviour and performance on a single mesh (here we chose a mesh of size $h = 0.0125$).

We first present the result for Algorithm 3.7 with Algorithm 3.10 which implies computing β, to the price of an additional solution in $\underline{u}$.

We shall use the conjugate gradient CG(n) method, preconditioned by SSOR for $\underline{u}$. For Algorithm 3.10 we use for $M_S = M_0$ and we illustrate its dependency to the solver in $\underline{u}$ by using different number of iterations in CG(n). We take the same tolerance as above.

In this case CG(3) yields the best performance (Table 4.5).

We now want to know if computing β is worth the cost. To do so, we take $CG(3)$ as the solver in $\underline{u}$ which was best in the previous test and we replace Algorithm 3.10 by Algorithm 3.11. This corresponds to the classical Arrow-Hurwicz method as described in [43]. Since β is arbitrary, we considered various values. Based on Table 4.6 the optimal value for a fixed β is between 8 and 9 and the corresponding number of iterations is about 400.

Table 4.6 Elasticity problem: Algorithm 3.7, number of iterations and CPU time according to the value of β

β	1	5	7	8	9
# iterations	Max iter	649	471	414	diverged
CPU time (s)	283	188	127	109	–

Table 4.7 Elasticity problem: performance of GCR's method preconditioned with Algorithm 3.7

Value of n (CG(n))	1	2	3	4	5
# iterations	117	91	82	98	81
CPU time (s)	29	30	33	47	51

One sees that the optimal β is slightly higher that 8 and that the computing time is the double than with a computed β. Moreover, the optimal values has to be guessed.

In the last comparison we use a GCR to accelerate the precedent solver, once again different number of iteration are taken for the CG(n).

In Table 4.7 the CPU time is increasing with n. However choosing a value of n between 1 and 3 achieves a good reduction of CPU time with respect to the optimal value of Table 4.5.

Effect of the Solver in $\underline{u}$

The next point that we will address is related to the use of HP solver. We are interested in the effect of HP-AMG and HP-LU when employed in the solver in $\underline{u}$.

In the next numerical tests we use a Mixed-GMP-GCR with different variants of Algorithm 3.7 with Algorithm 3.10 using solvers in $\underline{u}$ based on the HP-AMG or HP-LU.

We present in Fig. 4.5 the convergence of the residuals on the finer mesh for different solvers in $\underline{u}$ based on the HP-AMG.

One sees that the better the solver in $\underline{u}$, the better the convergence. Furthermore, the gain becomes negligible if this solver is good enough while using HP in PREONLY mode seems to be rather poor. However the picture is quite different if one considers computing time. This is what we present in Table 4.8.

For all three meshes, we present the computing time for the different versions of the mixed-GMP-GCR solver. Each row corresponds to the use of a specific version of the solver in $\underline{u}$. For coarse grids, the direct solver is much more efficient than the iterative ones but this advantage rapidly disappears when the mesh gets finer. The solvers using HP-AMG are clearly the best for large meshes. The good news is that there is little difference as soon as a good enough solver is employed.

Remark 4.12 (GCR or GMRES?) In the previous results, we have chosen GCR(n) for the solution in $\underline{u}$. One might wonder why this choice. Although mathematically

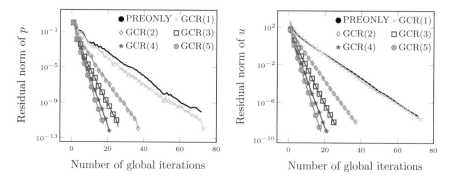

Fig. 4.5 Convergence of the residuals in p and $\underline{u}$ using a PREONLY(HP-AMG) or GCR(n)-HP-AMG with $n = 1, .., 5$

Table 4.8 Linear elasticity problem: CPU time in seconds according to the solver used for the primal problem and the size of the mesh. On top, as a reference, the CPU times using a direct solver (LU) for $\underline{u}$ at each iteration

		No. of degress of freedom		
		14,739	107,811	823,875
LU			7.09	70.41	1903.65
PREONLY	HP-AMG	11.30	49.43	253.11
	HP-LU	10.21	39.37	325.87
GCR(1)	HP-AMG	13.35	58.57	250.38
	HP-LU	11.39	50.92	350.62
GCR(2)	HP-AMG	10.10	46.92	179.11
	HP-LU	8.92	41.70	306.84
GCR(3)	HP-AMG	9.23	43.41	164.3
	HP-LU	7.92	39.93	313.00
GCR(4)	HP-AMG	8.44	38.08	159.78
	HP-LU	8.02	40.71	320.47
GCR(5)	HP-AMG	8.86	37.72	158.69
	HP-LU	8.20	42.41	353.74

GCR is equivalent to GMRES, this could indeed be discussed. To evaluate our choice, we have made a comparison with the GMRES(n). As we can see in Fig. 4.6, and as predicted, the behaviour is essentially the same, however for such saddle point system the GCR seems to benefit from orthogonality of the residuals as we observe a slight difference in the number of iterations. As to computing time, we still have a slight difference in favor of GCR as we see in Table 4.9 which should be compared with the last column of Table 4.8. ∎

All this shows that it is possible to solve correctly and rapidly incompressible linear elasticity problems, with a method which is totally automatic and does not rely on arbitrary parameters. We now consider the non linear case.

Fig. 4.6 Iterations number for the Mix-GMP-GCR preconditioned with a GCR(n)-HP or GMRES(n)-HP preconditioner

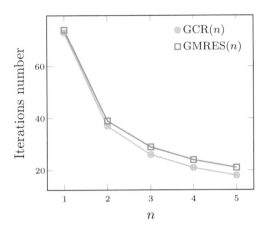

Table 4.9 Elasticity problem with a fine mesh (823 875 dof): CPU time according to the preconditionner of GMRES(n)-HP of the primal problem

	GMRES(n)				
	1	2	3	4	5
HP-AMG	32,163	22,051	21,581	21,381	19,980
HP-LU	50,314	39,616	48,920	33,563	50,829

4.2.6 Large Deformation Problems

In this section, we consider the solution of non linear elasticity problems for incompressible materials. We thus consider a Newton method which brings us to solve a sequence of linearised problems. We take as examples the neo-hookean and Money-Rivlin models. As we know, for non-linear elasticity, the coercivity of the linearised system (4.17) is not guaranteed. It might indeed fail near bifurcation points and this would require techniques [61] which are beyond the scope of our presentation. We shall focus on two points.

- The algorithms presented for the linear case are directly amenable to the linearised problems which we now consider.
- The stabilising terms are important.

In our numerical tests, we apply the Mixed-GMP-GCR method to the linearised problem. The solver in $\underline{u}$ is GCR(3) prconditioned by HP-AMG.

Neo-Hookean Material

The parameter of the considered material is $E = 10^2$ with a Poisson ratio $\nu = 0$ which gives $\mu = 50$. We regularize the problem with an artificial bulk modulus

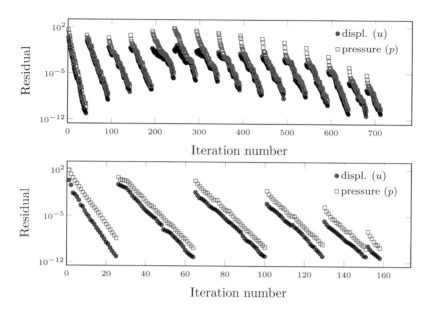

Fig. 4.7 Non linear elasticity problem (Neo-Hookean): convergence in l^2-norm of the residuals for Poisson ratio $\hat{\nu} = 0$ (left) and $\hat{\nu} = 0.4$ (right)

$\hat{k}$ corresponding to an artificial Poisson ratio $\hat{\nu}$. We can see in Fig. 4.7, how the stabilizing term accelerates the convergence of the problem.

In this result, GCR(3)-HP-AMG is used as the solver in $\underline{u}$. Each decreasing curve corresponds to a Newton iteration. Stabilisation does provide a better non linear behaviour. All the parameters of the algorithm are computed automatically.

Mooney-Rivlin Material

The material is rubber and the associated parameters are $c_1 = 0.7348$ et $c_2 = 0.08164$. With this material, the problem in $\underline{u}$ might not be coercive and as we solve at each Newton step a linearised problem the condition number will depend on the solution. In the formulation of (4.19), we introduced a stabilisation term $\hat{K}$. This stabilisation is indeed necessary.

The behaviour of the material for $\hat{K} = 0$ is somewhat paradoxical. When we solve in displacement only, we obtain the result on the left of Fig. 4.8. This kind of behaviour was also observed in the technical report of ADINA [2].

When we solve the incompressible problem with Mixed-GMP-GCR, we find a good solution, presented at the right of Fig. 4.8. However, we got a very slow convergence of the algorithm as we see in Fig. 4.9. For the first Newton's iteration all seems well but convergence is slow for the second one.

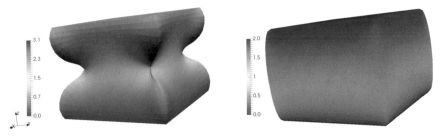

Fig. 4.8 Non linear elasticity problem (Mooney-Rivlin): solution of the problem for $\hat{K} = 0$. On the left using a displacement only formulation and on the right using a mixed formulation

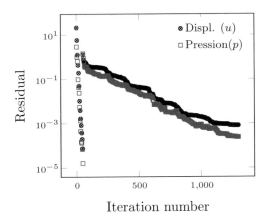

Fig. 4.9 Non linear elasticity problem in mixed formulation: convergence in l^2-norm of the residuals for $\hat{K} = 0$ for the two first Newton steps

When we stabilise with $\hat{K} > 0$ we have a better convergence (see Fig. 4.10). In this case an optimal value on $\hat{K}$ is 5 and is independent of the size of the problem. As we could expect, taking $\hat{K}$ too large has an adverse effect on the condition number of the problem in $\underline{u}$ and the algorithm slows down.

Finally we would like to confirm that the computed parameter β of Algorithm 3.10 is still correct for this non linear problem. In order to do so, we consider the same Mooney-Rivlin case as above using $\hat{K} = 10$ for stabilisation and tested with Algorithm 3.11 using some fixed values of β. Figure 4.11 allows us to compare the convergence for these fixed values of β with the behaviour when using β_{opt} (that is using Algorithm 3.10).

Obviously the choice of fixed values of β was not totally random, 'educated guess' were involved giving reasonable numerical behaviour. We can see again that the computed value of β is nearing the optimal value for this parameter. This, once again, justifies the use of Algorithm 3.10 and it also shows that the preconditioner is independent of the problem and avoids more or less justified guesses.

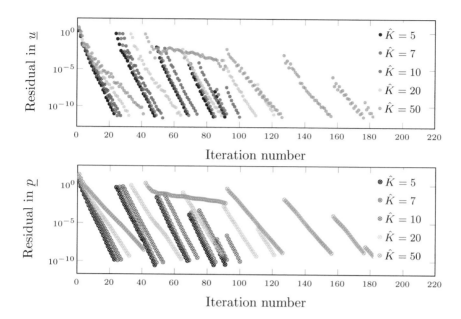

Fig. 4.10 Non linear elasticity problem (Mooney-Rivlin): convergence in l^2-norm of the residuals according to different artificial bulk modulus $\hat{K}$

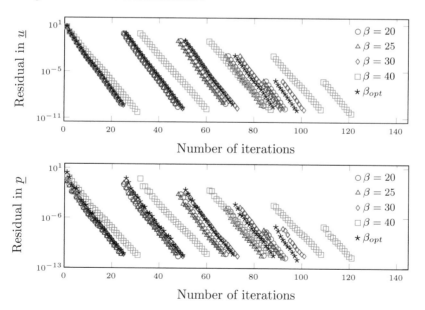

Fig. 4.11 Non linear elasticity problem (Mooney-Rivlin): convergence in l^2-norm of the residuals according to the value of β

4.3   Navier-Stokes Equations

We now rapidly consider Navier-Stokes equations for incompressible flow problems. This will evidently be very sketchy as numerical methods for Navier-Stokes equations is the subject of an immense literature. We consider the subject only as it is related to the previous Sect. 4.2 but also as it shares some points with Sect. 4.1 for the importance of the coercivity condition. This will lead us to present the classical projection method in a new setting. This should be seen as exploratory. Projection methods have been a major development in the solution methods for the Navier-Stokes equations, [29, 85]. We refer to [76] for a discussion of its variants and their relation to the type of boundary conditions imposed on the system.

We now present a simple form of the problem. Let Ω be a Lipschitzian domain in $\mathbb{R}^d$, $d = 2$ or 3. Let $\partial\Omega = \Gamma_D \cup \Gamma_N$ the boundary of Ω with $\Gamma_N \neq \emptyset$, $\underline{n}$ is the exterior normal vector and Ω_t the open set $\Omega \times (0, T)$, where $T > 0$ is the final time.

We consider an unsteady flow of an incompressible fluid in which the density variations are neglected and we restrict ourselves to a Newtonian fluid, that is with a constitutive equations of the form

$$\underline{\underline{\sigma}} = -p\mathbf{I} + 2\mu\underline{\underline{\varepsilon}}(\underline{u}),$$

where $\underline{\underline{\varepsilon}}(\underline{u})$ is defined as in (4.8). We consider homogeneous Dirichlet boundary conditions on Γ_D and a homogeneous Neumann condition on Γ_N,

$$\underline{u}(t, \underline{x}) = 0 \ \text{ on } \ \Gamma_D \qquad \underline{\underline{\sigma}}(\underline{u}(t, \underline{x}), p(t, \underline{x})) = 0 \ \text{ on } \ \Gamma_N.$$

We thus define,

$$V = \{\underline{v} \in (H^1(\Omega))^d | \ \underline{v} = 0 \ \text{ on } \ \Gamma_D\}, \ Q = L^2(\Omega),$$

$$a(\underline{u}, \underline{v}) = \int_\Omega \underline{\underline{\varepsilon}}(\underline{u}) : \underline{\underline{\varepsilon}}(\underline{v}) \, dx,$$

$$c(\underline{u}, \underline{v}, \underline{w}) = \int_\Omega \underline{u} \cdot \text{grad } \underline{v} \cdot \underline{w} \, dx.$$

We also denote

$$(\underline{u}, \underline{v}) = \int_\Omega \underline{u} \cdot \underline{v} \, dx$$

and we write the Navier-Stokes equations for the fluid velocity $\underline{u}$ and pressure p as

$$
\begin{cases}
(\partial \underline{u}/\partial t, \underline{v}) + c(\underline{u}, \underline{u}, \underline{v}) - 2\mu a(\underline{u}, \underline{v}) - (p, \operatorname{div} \underline{v}) = (\underline{f}, \underline{v}) & \forall \underline{v} \in V \\
(\operatorname{div} \underline{u}, q) = 0 & \forall q \in Q.
\end{cases}
\tag{4.24}
$$

Here $\underline{f}$ represents eventual external volumic forces. System (4.24) is completed with the following initial data:

$$
\underline{u}(0, \underline{x}) = \underline{u}_0(\underline{x}) \in L^2(\Omega)^d \text{ with } \operatorname{div} \underline{u}_0 = 0
$$

Here, we choose a uniform time step δt and a backward (also called implicit) Euler time discretisation. For the spatial discretisation we choose a finite element approximation (V_h, Q_h) for the velocity and pressure. At time $t^k = k\delta t < T$, knowing (u_h^{k-1}, p_h^{k-1}) we consider the system

$$
\begin{cases}
(\underline{u}_h^k, \underline{v}_h) + \delta t \, c(\underline{u}_h^k, \underline{u}_h^k, \underline{v}_h) - 2\mu \delta t \, a(\underline{u}_h^k, \underline{v}_h) \\
\qquad\qquad - \delta t(p_h^k. \operatorname{div} \underline{v}_h) = (\underline{u}_h^{k-1}, \underline{v}_h) + \delta t(\underline{f}, \underline{v}_h) & \forall \underline{v}_h \in V_h \\
(\operatorname{div} \underline{u}_h^k, q_h) = 0 & \forall q_h \in Q_h.
\end{cases}
\tag{4.25}
$$

Remark 4.13 We present this simple implicit time discretisation to fix ideas. Our development is in no way restricted to this example. In (4.25) we have a non linear problem for u_h^k for which we can consider a Newton linearisation. One could also consider a semi implicit formulation with $c(\underline{u}_h^{k-1}, \underline{u}_h^k, \underline{v}_h)$ instead of $c(\underline{u}_h^k, \underline{u}_h^k, \underline{v}_h)$. ∎

Let us denote $\widetilde{p} = \delta t \, p$. We can write (4.25) in the form,

$$
\begin{cases}
(\underline{u}_h^k, \underline{v}_h) - (\widetilde{p}_h^k. \operatorname{div} \underline{v}_h) + \delta t \, c(\underline{u}_h^k, \underline{u}_h^k, \underline{v}_h) - 2\mu \, \delta t \, a(\underline{u}_h^k, \underline{v}_h) \\
\qquad\qquad = (\underline{u}_h^{k-1}, \underline{v}_h) + \delta t(\underline{f}, \underline{v}_h) & \forall \underline{v}_h \in V_h \\
(\operatorname{div} \underline{u}_h^k, q_h) = 0 & \forall q_h \in Q_h.
\end{cases}
$$

We now apply he classical **projection method**. In its simplest form, we first compute a predictor $\widehat{\underline{u}}_h$ for u_h^k solution of,

$$
\begin{aligned}
(\widehat{\underline{u}}_h, \underline{v}_h) &+ \delta t \, c(\widehat{\underline{u}}_h, \widehat{\underline{u}}_h, \underline{v}_h) - 2\mu \, \delta t \, a(\widehat{\underline{u}}_h, \underline{v}_h) \\
&= (\underline{u}_h^{k-1}, \underline{v}_h) + (\widetilde{p}_h^{k-1}, \operatorname{div} \underline{v}_h) + \delta t(\underline{f}, \underline{v}_h) \quad \forall \underline{v}_h \in V_h
\end{aligned}
\tag{4.26}
$$

then project $\widehat{\underline{u}}_h$ on the divergence-free subspace by solving formally,

$$
\begin{cases}
-\Delta\delta\tilde{p} = \operatorname{div}\widehat{\underline{u}}_h \\[2mm]
\dfrac{\partial\delta\tilde{p}}{\partial n} = 0, \quad \text{on } \Gamma_D \\[2mm]
\delta\tilde{p} = 0, \quad \text{on } \Gamma_N
\end{cases}
\tag{4.27}
$$

and finally updating the velocity and pressure

$$
\begin{cases}
\underline{u}_h^k = \widehat{\underline{u}}_h - \operatorname{grad}\delta\tilde{p} \\[2mm]
\tilde{p}^k = \tilde{p}^{k-1} + \delta\tilde{p}
\end{cases}
$$

The basic flaw in this approach is that this projection takes place in $H(\operatorname{div}, \Omega)$ and not in $H^1(\Omega)$. Tangential boundary values are lost. Many possibilities have been explores to cure this and we shall propose one below.

In a finite element formulation, the exact meaning of $\widehat{\underline{u}}_h - \operatorname{grad}\delta p$ must also be precised. We consider a non standard way of doing this.

Referring to our mixed formulation (4.3) and defining $\delta\underline{u}_h = \underline{u}_h^k - \widehat{\underline{u}}_h$

$$
V_{0h} = \{v_h \mid \underline{v}_h \cdot \underline{n} = 0 \text{ on } \gamma_D\}.
\tag{4.28}
$$

The Poisson problem (4.27) can be written as finding $\delta\underline{u}_h \in V_{0h}$, solution of

$$
\begin{cases}
(\delta\underline{u}_h, \underline{v}_h) + (\delta\tilde{p}_h. \operatorname{div}\underline{v}_h) = 0 & \forall\underline{v}_h \in V_{0h} \\[2mm]
(\operatorname{div}\delta\underline{u}_h, q_h) = (\operatorname{div}\widehat{\underline{u}}_h, q_h) & \forall q_h \in Q_h.
\end{cases}
\tag{4.29}
$$

We have an incompressibility condition and we have to choose a discrete formulation satisfying the inf-sup condition. To fix ideas, we may assume the same choice as we used earlier for incompressible elasticity (cf. Sect. 4.2.3), that is the Taylor-Hood element which is also classical for flow problems. It is not, however, suitable for the formulation (4.29). The trouble is with the lack of coercivity on the kernel. This has been studied in [25] and the cure has already been introduced : we add in (4.29) a stabilising term to the first equation and we now have to find $\delta\underline{u}_h \in V_{0h}$ solution of

$$
\begin{cases}
(\delta\underline{u}_h, \underline{v}_h) + \alpha(\operatorname{div}\delta\underline{u}_h - \operatorname{div}\widehat{\underline{u}}_h, \operatorname{div}\underline{v}_h) \\[2mm]
\qquad\qquad + (\delta\tilde{p}_h. \operatorname{div}\underline{v}_h) = 0 & \forall\underline{v}_h \in V_{0h} \\[2mm]
(\operatorname{div}\delta\underline{u}_h, q_h) = (\operatorname{div}\widehat{\underline{u}}_h, q_h) & \forall q_h \in Q_h.
\end{cases}
\tag{4.30}
$$

With the discretisation considered, this will not yield a fully consistent augmented Lagrangian and the stabilising parameter will have to be kept small. We are fortunate: our experiments of Sect. 4.1 show that even a small value will be enough to get a good convergence.

Remark 4.14 (Why This Strange Mixed Form) Using the mixed form (4.30) makes $\widehat{u}_h$ and grad p_h to be in the same finite element space, which is a nice property.

We recall that the normal condition $\delta u_h \cdot n = 0$ must be imposed explicitly on Γ_D in either (4.29) or (4.30). If no condition is imposed on $\delta u_h \cdot n$, one then imposes $\delta \tilde{p}_h = 0$ in a weak form.

■

Since we have no control on the tangential part of grad p_h on Γ_D this method potentially leaves us with a tangential boundary conditions which is not satisfied. The simplest remedy would be an iteration on the projection. To fix ideas, we shall use the semi-implicit problem

$$
\begin{cases}
(u_h^k, v_h) + \delta t \, c(u_h^{k-1}, u_h^k, v_h) - 2\mu \delta t \, a(u_h^k, v_h) \\
\qquad - (\tilde{p}_h^k \cdot \operatorname{div} v_h) = (u_h^{k-1}, v_h) + \delta t(f, v_h) \qquad \forall v_h \in V_h \\
(\operatorname{div} u_h^k, q_h) = 0 \qquad \forall q_h \in Q_h.
\end{cases} \qquad (4.31)
$$

One could then, given an initial $\tilde{p}_h^k$,

1. Solve the first equation of (4.31)
2. Project the solution by (4.30).
3. Update $\tilde{p}_h^k + \delta \tilde{p}_h$.
4. Repeat until convergence.

From this, one can think of many possibilities using approximate solutions and imbedding the procedure, for example, in a GCR method. In all cases, solving (4.30) will require a mixed iteration. and this leads us to another approach.

4.3.1 A Direct Iteration: Regularising the Problem

Finally we consider the direct use of the same Algorithms 3.7 and Algorithm 3.10 that we used for incompressible elasticity. To get a better coercivity we had the regularised penalty terms and we change the first equation of (4.25) into

$$
(u_h^k, v_h) + \alpha(\operatorname{div} u_h^k, \operatorname{div} v_h) + \delta t \, c(u_h^k, u_h^k, v_h) - 2\mu \delta t \, a(u_h^k, v_h)
$$
$$
- \delta t(p_h^k \cdot \operatorname{div} v_h) = (u_h^{k-1}, v_h) + \delta t(f, v_h) \qquad \forall v_h \in V_h \qquad (4.32)
$$

This is a non linear problem which should be linearised. We can then apply the Mixed-GMP-GCR method to the resulting linearised form.

4.3.2 A Toy Problem

In order to show that this technique is feasible, we consider a very simple example. We consider $\Omega =]0, 1[\times]0, 1[$ and an artificial solution,

$$\underline{u}(t, x, y) = (y^2(1 + t), x^2(1 + t))$$

with a viscosity's value equal to 10^{-2}, the source force associated is

$$f(t, x, y) = \big(0.98(1 + t) + y^2 + 2x^2(1 + t)^2 y,$$
$$- 1.02(1 + t) + x^2 + 2y^2(1 + t)^2 x\big)$$

Results for the method (4.32) are presented in Table 4.10, which represents the global number of iterations and CPU time in seconds to reach $T = 1$. This is done for different values of the time step δt and the regularity coefficient α. The table shows the interest of the stabilisation parameter α, which can reduce the number of global iterations and CPU time. The optimal value of the stabilisation parameter is also stable with respect to the time step. We used an iterative solver (precisely CG(10)) for solving the problem in $\underline{u}$, a choice which could clearly be improved. The mesh is composed of 3200 triangles and 6561 degrees of freedom.

The following figures show the convergence for the problem with $\alpha = 0.1$ and $\delta t = 10^{-1}$. Figure 4.12 presents the convergence of both primal and dual residuals for Newton iterations of a single arbitrary time step (here the fifth one corresponding to $t = 0.5$).

Table 4.10 Navier-Stokes problem: iteration's number and CPU time in second according to the step time (δt) and the regularity coefficient (α)

α	$\delta t = 0.05$		$\delta t = 0.1$		$\delta t = 0.25$	
	# it.	CPU (s)	# it.	CPU (s)	# it.	CPU (s)
0	13,593	945	6701	475	3542	254
1×10^{-2}	10,567	739	5167	366	2765	178
1×10^{-1}	5709	412	3203	233	2231	161
2.5×10^{-1}	5440	396	3756	267	3150	221
1	8315	580	6767	468	6033	441

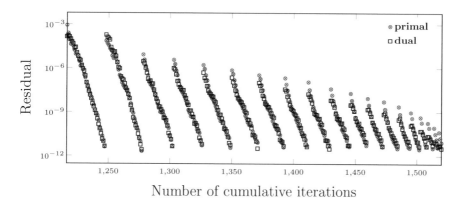

Fig. 4.12 Navier-Stokes problem: convergence curves for the primal and dual residuals for $t = 0.5$

The same behaviour repeats at every time step.

This is a very simple case. Nevertheless, we may think that this direct approach is feasible. Clearly, a better solver in u is needed and it is thinkable, following [26] to include the projection method in the preconditioner.

Chapter 5
Contact Problems: A Case Where $Q \neq Q'$

This chapter presents solution methods for the sliding contact. We shall first develop some issues related to functional spaces: indeed, in contact problems, we have a case where the space of multipliers is not identified to its dual. To address this, we first consider the case where a Dirichlet boundary condition is imposed by a Lagrange multiplier and present the classical obstacle problem as a simplified model for the contact problem. We shall then give a description of the contact problem and its discretisation with a numerical example.

5.1   Imposing Dirichlet's Condition Through a Multiplier

We temporarily put aside contact problem to consider a simple Dirichlet problem which will nevertheless enable us to address some issues relative to contact problems.

We thus consider a domain Ω of boundary Γ. We divide Γ into two parts Γ_0 where null Dirichlet are imposed and Γ_C where we have a non homogeneous Dirichlet condition which will provide a simplified model of the contact region. We then want to solve

$$\begin{cases} -\Delta u & = f \quad \text{in} \Omega \\ u & = 0 \quad \text{on} \Gamma_0 \\ u & = g \quad \text{on} \Gamma_C \end{cases} \qquad (5.1)$$

We define the space $V = \{v \mid v \in H^1(\Omega),\ v|_{\Gamma_0} = 0\}$. We suppose that $f \in V'$ and $g \in H_{00}^{1/2}(\Gamma_D)$. Following Sect. 2.1.5, we take $\Lambda = H_{00}^{1/2}(\Gamma_C)$ and the operator $\mathcal{B}$ is the trace of V in Λ on Γ_C. One should note that we take $g \in H_{00}^{1/2}(\Gamma_C)$ to make our presentation simpler but that this is not essential.

© The Author(s), under exclusive license to Springer Nature Switzerland AG 2022
J. Deteix et al., *Numerical Methods for Mixed Finite Element Problems*,
Lecture Notes in Mathematics 2318, https://doi.org/10.1007/978-3-031-12616-1_5

Remark 5.1 (Change of Notation) In the following, we denote Λ instead of Q the space of multipliers. ∎

Remark 5.2 (Sobolev Spaces) The space $H_{00}^{1/2}(\Gamma_C)$ was introduced in [62]. The elements of this space are in a weak sense null at the boundary of Γ_C and thus match with the zero boundary conditions on Γ_0. The dual of $H_{00}^{1/2}(\Gamma_C)$ is $H^{-1/2}(\Gamma_C)$. The scalar product on $H_{00}^{1/2}(\Gamma_C)$ is usually defined by an interpolation norm which also defines a Ritz operator $\mathcal{R}$ from $H_{00}^{1/2}(\Gamma_C)$ onto $H^{-1/2}(\Gamma_C)$. We shall consider later other possibilities to define the scalar product. ∎

Let then $\mathcal{B}v$ be the trace of v in $H_{00}^{1/2}(\Gamma_C)$. To write our problem in mixed form, we define

$$a(u, v) = \int_{\Omega} \text{grad } u \cdot \text{grad } v \, dx \qquad (5.2)$$

$$b(v, \mu) = (\mathcal{B}v, \mu)_{H_{00}^{1/2}(\Gamma_C)} = \langle \mathcal{B}v, \mathcal{R}\lambda \rangle_{H_{00}^{1/2}(\Gamma_C) \times H^{-1/2}(\Gamma_C)}$$

To simplify the notation we write $(\lambda, \mu)_{1/2}$ the scalar product in $H_{00}^{1/2}$. We want to find $u \in V, \lambda \in \Lambda$ solution of

$$\begin{cases} a(u, v) + b(v, \lambda) = (f, v) & \forall v \in V, \\ b(u, \mu) = (g, \mu)_{1/2} & \forall \mu \in \Lambda. \end{cases} \qquad (5.3)$$

This problem is well posed. Indeed the bilinear form $a(u, v)$ is coercive and we have an inf-sup condition. To show this, we use the fact that there exists a continuous lifting $\mathcal{L}$ from $H_{00}^{1/2}(\Gamma_C)$ into V. Denoting $v_\lambda = \mathcal{L}\lambda$, we have

$$|v_\lambda|_V \leq C|\lambda|_\Lambda$$

$$\sup_{v} \frac{b(v, \lambda)}{|v|_V} \geq \frac{b(v_\lambda, \lambda)}{|v_\lambda|_V} \geq \frac{1}{C}|\lambda|_\lambda.$$

Remark 5.3 (This May Seem a Non Standard Formulation!) Taking $\lambda \in H_{00}^{1/2}(\Gamma_C)$ may seem a little strange. Indeed, a more standard formulation would define

$$b(v, \mu') = \langle \mathcal{B}v, \mu' \rangle_{H_{00}^{1/2}(\Gamma_C) \times H^{-1/2}(\Gamma_C)}$$

In fact the two formulations are equivalent as we have

$$\langle \mathcal{B}u, \mu' \rangle_{H_{00}^{1/2}(\Gamma_C) \times H^{-1/2}(\Gamma_C)} = (\mathcal{B}u, \mathcal{R}^{-1}\mu')_{1/2}.$$

where $\mathcal{R}$ is the Ritz operator on Λ. We therefore have the choice of working with λ or with λ', and this choice will be dictated by numerical considerations. ∎

We may then introduce the operator $B = \mathcal{R}\mathcal{B}$ from V onto Λ' and we have

$$\langle Bu, \lambda \rangle_{\Lambda' \times \Lambda} = b(v, \lambda)$$

Remark 5.4 We have considered a case where the choice of Λ and Λ' is simple. In more realistic situations Λ is a subspace of $H^{1/2}(\Gamma_C)$ corresponding to the trace of the elements of a subspace of $H^1(\Omega)$. The space Λ' is then the dual of Λ which in general will contain boundary terms. ∎

As a simple numerical procedure to solve (5.3) we could use Algorithm 3.7. A central point of this algorithm would be to compute,

$$z_\lambda = \tilde{S}^{-1} r_\lambda = \tilde{S}^{-1}(Bu - g).$$

It will thus be important to have a good approximation of the operator S^{-1}. To better understand the issue, we have to analyse more closely the spaces considered here.

5.1.1 $H_{00}^{1/2}(\Gamma_C)$ and its dual $H^{-1/2}(\Gamma_C)$

We rapidly present some results that should help to understand the spaces with which we have to work. In the present case, $\mathcal{R}$ is the Ritz operator from Λ into Λ', that is from $H_{00}^{1/2}(\Gamma_D)$ onto $H^{-1/2}(\Gamma_D)$. This corresponds to the Dirichlet-Neumann Steklov-Poncaré operator which we shall consider in more detail below.

The classical way to define $H_{00}^{1/2}(\Gamma_C)$ is through an interpolation between L^2 and H_0^1. There has been in the last years many developments to define fractional order Sobolev spaces in term of fractional order derivatives. We refer to [31] for a general presentation. We show informally how this could bring some insight about our problem. If we suppose to simplify that Γ_C lies in a plane, the elements of $H_{00}^{1/2}(\Gamma_C)$ can be prolonged by zero as elements of $H_{00}^{1/2}(\mathbb{R}^{n-1})$ and we can then introduce on Γ_C a fractional order gradient of order $1/2$ denoted $\underline{\mathrm{grad}}^{\,1/2}$. It must be noted that fractional derivatives are not local operators.

We can define for $\Omega \subset \mathbb{R}^n$ and $\Gamma_C \subset \mathbb{R}^{n-1}$

$$H_{00}^{1/2}(\Gamma_C) = \{v \mid v \in L^2(\Gamma_C), \ \underline{\mathrm{grad}}^{\,1/2} v \in (L^2(\Gamma_C))^{n-1}\}$$

We can also define on $H_{00}^{1/2}(\Gamma_C)$ a norm

$$|v|_{1/2}^2 = \int_{\Gamma_C} |v|^2 dx + \int_{\Gamma_C} |\underline{\mathrm{grad}}^{\,1/2} v|^2 \, dx \tag{5.4}$$

This norm depends on a fractional tangential derivative.

Remark 5.5 (Discrete Interpolation Norms) This should be made in relation to discrete interpolation norms [5], which involve taking the square root of the matrix associated to a discrete Laplace-Beltrami operator on the boundary. This is an operation which might become quite costly for large problems. One should also recall the work of [44]. ∎

We can then identify the dual of $H_{00}^{1/2}$ with $L^2(\Gamma_C) \times (L^2(\Gamma_C))^{n-1}$ and the duality product as

$$\langle \lambda, v \rangle = \int_{\Gamma_C} \lambda_0 \, v dx + \int_{\Gamma_C} \underline{\lambda}_1 \cdot \underline{\text{grad}}^{1/2} v \, dx$$

with $\lambda_0 \in L^2(\Gamma_C)$ and $\underline{\lambda}_1 \in (L^2(\Gamma_C))^{n-1}$. As regular functions are dense in $H_{00}^{1/2}(\Gamma_C)$, we could say that $H^{-1/2}(\Gamma_C)$ is the sum

$$\lambda_0 + \text{div}^{1/2} \underline{\lambda}_1$$

where $\text{div}^{1/2}$ is a fractional order divergence operator.

5.1.2 A Steklov-Poincaré operator

Another important component of our problem is the Dirichlet to Neumann Steklov-Poincaré operator. Essentially, the idea is to associate to a Dirichlet condition the corresponding Neumann condition. This is a wide subject and we present the simple case associated with our model problem. We refer to [77] for the discussion of domain decomposition methods.

Given $r \in H_{00}^{1/2}(\Gamma_C)$, we first use the continuous lifting from $H_{00}^{1/2}(\Gamma_C)$ into V and find u_r such that $\mathcal{B}u_r = r$ on Γ_C. We then solve for $u_0 \in H_0^1(\Omega)$

$$a(u_0, v_0) = -a(u_r, v_0), \quad \forall v_0 \in H_0^1(\Omega).$$

and setting $\phi_r = u_0 + u_r$, we can then define $\lambda' \in H^{-1/2}(\Gamma_C)$ by

$$\langle \lambda', \mathcal{B}v \rangle_{\Lambda' \times \Lambda} = a(\phi_r, v) \quad \forall v \in H^1(\Omega)$$

Remark 5.6 Taking $v = u$, $\langle \lambda', \mathcal{B}u \rangle_{\Lambda' \times \Lambda}$ is a norm on $H_{00}^{1/2}(\Gamma_C)$. ∎

Using This as a Solver

To solve the Dirichlet problem we can now build a Neumann problem using the $\mathcal{SP}$ operator. Assuming we have some initial guess λ_0' we shall first solve the Neumann problem

$$a(u_N, v) = \langle f, v \rangle + \langle v, \lambda_0' \rangle_{\Lambda \times \Lambda'} \quad \forall v \in V.$$

We then set $r = g - \mathcal{B}u_N$ on Γ_C, and get $\lambda' = \lambda_0' + \mathcal{SP}r$.

Solving the Dirichlet problem (5.1) is then equivalent to solving the Neumann problem

$$a(u, v) + \langle \mathcal{B}v, \lambda' \rangle_{\Lambda \times \Lambda'} = \langle f, v \rangle \quad \forall v \in V. \tag{5.5}$$

The condition $\mathcal{B}u = g$ on Γ_C can be written as $(\mathcal{B}u - g, \mathcal{B}v)_\Lambda = 0$ which can also be written as

$$\langle \mathcal{B}u - g, \mathcal{SP}\,v \rangle_{\Lambda \times \Lambda'} = \langle \mathcal{B}u - g, \mu' \rangle_{\Lambda \times \Lambda'} = 0 \quad \forall \mu' \in \Lambda'. \tag{5.6}$$

Remark 5.7 We have thus written our problem in the form (5.3). One could argue that we did nothing as the Steklov-Poincaré operator implies the solution of a Dirichlet's problem. This will become useful whenever the λ' has an importance by itself. This will be the case in a similar formulation of the contact problem where λ is the physically important contact pressure.   ∎

5.1.3   Discrete Problems

The direct way of discretising problem (5.3) would be to consider a space $V_h \subset V$ and to take for Λ_h a subspace of the space of the traces of V_h on Γ_C. We then have $\mathcal{B}_h v_n = P_{\Lambda_h} \mathcal{B}v_h$. We define

$$b(v_h, \lambda_h) = (\mathcal{B}_h v_h, \lambda_h)_{1/2, h},$$

where $(\lambda_h, \mu_h)_{1/2, h}$ is some discrete norm in Λ_h. We look for u_h and λ_h solution of

$$\begin{cases} a(u_h, v_h) + b(v_h, \lambda_h) = (f, v_h) & \forall v_h \in V_h, \\ b(u_h, \mu_h) = (g_h, \mu_h)_{1/2, h} & \forall \mu_h \in \Lambda_h, \end{cases} \tag{5.7}$$

where we may take g_h the interpolate of g in Λ_h. More generally g_h might be the projection of g on Λ_h in $(\cdot, \cdot)_{1/2, h}$ norm.

The Matrix Form and the Discrete Schur Complement

The scalar product defines a discrete Ritz operator $\mathcal{R}_h$ from Λ_h onto Λ'_h. We have,

$$(\mathcal{B}_h v_h, \mu_h)_{1/2,h} = \langle \mathcal{R}_h \mathcal{B}_h v_h, \mu_h \rangle_{\Lambda'_h \times \Lambda_h} = \langle \mathcal{B}_h v_h, \mu_h \rangle_{\Lambda'_h \times \Lambda_h}$$

In the framework of Sect. 2.2.2 and Remark 2.5 we can associate to u_h and λ_h their coordinates u and λ on given bases.

R will the matrix associated to $\mathcal{R}_h$ that is

$$\langle R\lambda, \mu \rangle = (\lambda_h, \mu_h)_{1/2,h}$$

We then define the matrix B

$$\langle B\, u, \lambda \rangle = (\mathcal{B}_h u_h, \lambda_h)_{1/2,h}$$

- We emphasise that these matrices depend on the discrete scalar product.

  The Schur complement $BA^{-1}B^t$ operates from Λ_h onto λ'_h and can be read as

1. Solve the Neumann problem $a(u_h, v_h) = (\lambda_h, \mathcal{B}_h v_h)_{1/2,h}$.
2. Take the trace $\mathcal{B}_h u_h$ and compute $\lambda'_h = \mathcal{R}_h \mathcal{B}_h u_h$.   ∎

   This is usable if the bilinear form $b(v_h, \lambda_h)$, that is the discrete scalar product in $H_{00}^{1/2}$ is easily computable.

Remark 5.8 We can also write the discrete problem using the discrete form of (5.5) and (5.6). We then look for $u_h \in V_h$ and λ'_h in Λ'_h solution of

$$\begin{cases} a(u_h, v_h) + \langle v_h, \lambda'_h \rangle_{\Lambda_h \times \Lambda'_h} = (f, v_h) & \forall v_h \in V_h, \\ \langle \mathcal{B}_h u_h - g_h, \mu'_h \rangle_{\Lambda_h \times \Lambda'_h} = 0 & \forall \mu'_h \in \Lambda'_h. \end{cases}$$

As in the continuous case, we can write the problem using $\Lambda_h \in \Lambda_h$ or $\lambda'_h \in \Lambda'_h$. The two formulations are equivalent for the equality condition $u = g$ but this will not be the case for inequality condition $u \geq g$.   ∎

We shall first rely, on a discrete Steklov-Poncaré operator which will enable us to define a 'perfect' discrete scalar product in Λ_h.

5.1.4 A Discrete Steklov-Poincaré Operator

We thus take as Λ_h a subspace of the traces of functions of V_h on Γ_C and we write $\mathcal{B}_h v_h$ the projection of $\mathcal{B} v_h$ on Λ_h.

We shall also denote the matrix B associated to $\mathcal{B}_h$., the same notation as the continuous operator $!B$ as they are used in a different context.

The goal is to associate to an element $r_h \in \Lambda_h$ an element λ'_h in Λ'_h. To do so, we first build ϕ_{hr} a function of V_h such that $\mathcal{B}_h \phi_{hr} = r_h$ and we solve

$$a(\phi_{h0}, v_{h0}) = -a(\phi_{hr}, v_{h0}) \quad \forall v_{h0}$$

where $v_{h0} = 0$ on Γ_C and the bilinear form is as in (5.2). Let $\phi_h = \phi_{h0} + \phi_{hr}$. We now define $\mathcal{SP}_h r_h = \lambda'_h$, an element of Λ'_h by

$$\langle \mathcal{B}_h v_h, \lambda'_h \rangle_{\Lambda_h \times \Lambda'_h} = a(\phi_h, v_h) \quad \forall v_h \in V_h. \tag{5.8}$$

This would enable us to solve the problem (5.7) just as we had done for the continuous case:

- Given λ'_{0h} solve the Neumann problem

$$a(\hat{u}_h, v_h) = (f, v_h) + \langle v_h, \lambda'_{0h} \rangle \quad \forall v_h \in V_h.$$

- Take $r_h = \mathcal{B} u_h - g_h$ and compute $\lambda'_h = \lambda'_{0h} + \mathcal{SP}_h r_h$
- Solve

$$a(u_h, v_h) = (f, v_h) + \langle \lambda'_h, v_h \rangle \quad \forall v_h \in V_h. \tag{5.9}$$

What we have done here is to define the discrete scalar

$$(\lambda_h, \mu_h)_{1/2,h} = \langle \mathcal{SP}_h \lambda_h, \mu_h \rangle_{\Lambda'_h \times \Lambda_h}.$$

Again, this is somehow tautological: we solve a Dirichlet problem by solving a Dirichlet problem. The gain is that we also have λ'_h which has in contact problems a physical significance. If we write u and λ' the vectors asociated with u_h and λ'_h, we can introduce a matrix SP and write (5.1.4) as

$$(\lambda_h, \mu_h)_{1/2,h} = \langle SP\lambda, \mu \rangle. \tag{5.10}$$

We shall consider later simplified formulations. It is worth, however to give a look at the computation of the discrete Steklov-Poincaré operator.

5.1.5 Computational Issues, Approximate Scalar Product

We first note that to compute $\mathcal{SP}_h r_h$ as in (5.8), we need only to compute $a(\phi_h, v_{ih})$ for all v_{ih} associated to a node i on Γ_C as in Fig. 5.1.

Fig. 5.1 Node of the boundary

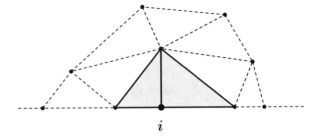

i

It is also interesting to see what this result looks like. To fix ideas, let us consider a piecewise linear approximation for V_h.

Referring to Fig. 5.1, we obtain at node i

$$a(\phi, v_{ih}) = h_x h_y[(\underline{\mathrm{grad}}_y \, \phi \, (-1/h_y) + \underline{\mathrm{grad}}_x \, \phi \, \underline{\mathrm{grad}}_x \, v_{ih}]. \tag{5.11}$$

The first term can be read as

$$h_x \, \underline{\mathrm{grad}}_y \, \phi \approx \int_{\Gamma_C} \frac{\partial \phi}{\partial n} v \, ds.$$

But this is not all: the second term depends on the tangential derivative of ϕ_r near the boundary, and could be thought as a half order derivative of the boundary value r. We have

$$h_y \frac{(r_{i+1} - 2r_i + r_{i-1})}{h_x}$$

It is interesting to compare this to (5.4). This suggests to define a discrete scalar product to approximate the $H_{00}^{1/2}(\Gamma_C)$ scalar product

$$(\lambda_h, \mu_h)_h = \int_{\Gamma_C} \lambda_h \, \mu_h ds + h \int_{\Gamma_C} \underline{\mathrm{grad}} \, \lambda_h \, \underline{\mathrm{grad}} \, \mu_h ds. \tag{5.12}$$

Formulas more or less similar to this are often advocated.   ∎

Remark 5.9 (A Matrix Representation of $\mathcal{SP}_h$)

In Sect. 5.1.4 we have introduced a matrix representing $\mathcal{SP}_h$ We could also define this matrix by computing ϕ_{ih} for every u_{ih} on Γ_C, computing

$$SP_{ij} = a(\phi_{ih}, \phi_{jh})$$

As noted above this reduces to a computation near the boundary.

There is a cost to this: one must solve a Dirichlet problem for every nodal value. Whether this is worth the effort would depend on each problem and the number of cases where this matrix form would be used and the cleverness with which we solve the Dirichlet problems.   ∎

Simplified Forms of the $\mathcal{SP}_h$ Operator and Preconditioning

We shall ultimately solve our discrete problem (5.7) by Algorithm 3.7 with a suitable preconditioner. In Algorithm 3.10, our standard preconditioner, we compute an approximation of the Schur complement $B\tilde{A}^{-1}B^t$ and approximate its inverse by M_S. In this case, the iteration is don in λ.

One can see $\mathcal{SP}_h$ as a representation of the Schur complement. To employ it in Algorithm 3.7, one could rely on an approximation $\widetilde{\mathcal{SP}}_h$

- One could think of building $\widetilde{\mathcal{SP}}_h$ on a subdomain around Γ_C and not on the whole Ω.
- The computation of $\widetilde{\mathcal{SP}}_h$ could also be done using a simpler problem, for instance using a Laplace operator instead of an elasticity operator.
- The Dirichlet problem defining $\widetilde{\mathcal{SP}}_h$ could be solved only approximately with $\tilde{A}$.

One should now modify Algorithm 3.10 as we are now iterating in λ' and not in λ. We must also note that we have an approximation $\widetilde{SP}$ of the Schur complement S.

Algorithm 5.1 *A preconditioner using $\widetilde{\mathcal{SP}}_h$*

1: Initialization

- r_u, r_λ *given ,*
- $z_u = \tilde{A}^{-1} r_u$
- $\tilde{r}_u = r_u + \mathcal{B} z_u$
- $z'_\lambda = \widetilde{SP} \tilde{r}_u$

- $zz_u = \tilde{A}^{-1} = B^t(\lambda')$
- $T_u = \mathcal{B} zz_u$
- $\beta = \dfrac{\langle z'_\lambda, T_u \rangle}{\langle T_u, T_u \rangle}$

2: Final computation

- $z'_\lambda = \beta z'_\lambda$
- $z_u = z_u - \beta zz_u$

3: End   ∎

The real efficiency of using a $\mathcal{SP}_h$ operator is to be tested. We leave this as an open point.

5.1.6 The $L^2(\Gamma_C)$ Formulation

A very common choice is to take as a scalar product

$$(\lambda_h, \mu_h)_{1/2,h} = \int_{\Gamma_C} \lambda_h \, \mu_h ds \qquad (5.13)$$

In this formulation, although $\lambda_h \in L^2(\Gamma_C)$, it is in fact a representation of λ_h'. We must expect a weak convergence in $H^{-1/2}(\Gamma_C)$.

Formally, this also means that, using (5.13) we shall have an inf-sup condition (2.19) in L^2 with $\beta_h = O(h^{1/2})$.

The Choice of Λ_h

We have considered the case where Λ_h is the trace of V_h. We could also take a subspace of the traces. A simple and important example would be to have piecewise quadratic elements for V_h and a piecewise linear subspace for Λ_h. We refer to [24] for an analysis of the inf-sup condition and the choice of spaces. Why would we do this? Essentially because of β in the inf-sup condition: A larger Λ_h means a smaller β_h. This is a standard point in the analysis of mixed methods, augmenting the space of multipliers makes the inf-sup condition harder to satisfy. We then have two consequences.

- As λ_h is a representation of λ_h' which converges in $H^{-1/2}$, a richer Λ_h' will produce an oscillatory looking λ_h.
- A smaller β will mean a slower convergence of the solver.

If we take a reduced Λ_h it must be noted that the solution is changed as $\mathcal{B}_h u_h = g_h$ means

$$P_{\Lambda_h}(\mathcal{B}u_h - g_h) = 0$$

which is weaker than $\mathcal{B}u_h = g_h$.

5.1.7 A Toy Model for the Contact Problem

Introducing a multiplier to manage a Dirichlet condition was a first step in exploring some of the technicalities related to contact problems. Another issue will arise when considering contact problems. We shall have to introduce positivity constraints. To illustrate this, we consider an **obstacle problem** in which the Dirichlet condition

$$\mathcal{B}u = g \quad \text{on } \Gamma_C$$

is replaced by an inequation $\mathcal{B}u \geq g$ on Γ_C. For the continuous problem, we use $\Lambda = H_{00}^{1/2}(\Gamma_C)$ and

$$b(v, \mu) = (\mathcal{B}v, \mu)_{1/2} = \langle \mathcal{R}\mathcal{B}v, \mu \rangle_{\Lambda' \times \lambda} = \langle \mathcal{B}v, \mu \rangle_{\Lambda' \times \lambda}.$$

We can then consider in Λ, the cone Λ^+ of almost everywhere positive μ^+ and on Λ'

$$\Lambda^{+'} = \{\mu^{+'} \mid \langle \mu^{+'}, \mu^+ \rangle_{\Lambda' \times \lambda} \geq 0 \; \forall \mu^+ \in \Lambda^+\} \tag{5.14}$$

We thus solve for $u \in V$, $\lambda^+ \in \Lambda^+$ solution of

$$\begin{cases} a(u, v) + b(v, \lambda^+) = (f, v) & \forall v \in V, \\ b(u, \mu^+ - \lambda^+) \geq (g, \mu^+ - \lambda^+)_{1/2} & \forall \mu^+ \in \Lambda^+. \end{cases}$$

The solution satisfies the Kuhn-Tucker conditions.

$$\lambda^+ \in \Lambda^+, \quad (\mathcal{B}u - g, \mu^+)_{1/2} \geq 0 \quad \forall \mu^+ \in \Lambda^+, \quad (\mathcal{B}u - g, \lambda^+)_{1/2} = 0.$$

The second condition can be read as

$$\langle \mathcal{R}(\mathcal{B}u - g), \mu^+ \rangle_{\Lambda' \times \lambda} \geq 0 \quad \forall \mu^+ \tag{5.15}$$

We have thus imposed the weak condition

$$(\mathcal{B}u - \mathcal{R}g) \geq 0 \text{ in } \Lambda'. \tag{5.16}$$

The Discrete Formulation

To get a discrete version, we have to choose V_h and Λ_h and we must give a sense to $\lambda_h \geq 0$.

For the choice of V_h, we take a standard approximation of $H^1(\Omega)$. For Λ_h we take the traces of V_h or more generally a subspace. For example, if V_h is made of quadratic functions, we can take Λ_h the piecewise quadratic traces or a piecewise linear subspace.

To define Λ_h^+ the obvious choice is to ask for nodal values to be positive. As we can see in Fig. 5.2 this works well for piecewise linear approximation. For a piecewise quadratic approximation one sees that positivity at the nodes does not yield positivity everywhere. Piecewise linear approximations are thus more attractive even if this is not mandatory.

Fig. 5.2 Positivity of λ

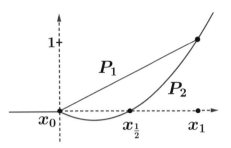

As in the continuous case, the definition of positivity on Λ_h also defines positivity on Λ'_h

$$\lambda'_h \geq 0 \iff \langle \lambda'_h, \mu_h \rangle_{\Lambda'_h \times \Lambda_h} \geq 0 \quad \forall \mu_h \geq 0$$

Let $(\lambda_h, \mu_h)_{1/2,h}$ be a scalar product on Λ_h. We then define $b(v_h, \mu_h) = (\mathcal{B}_h v_h, \mu_h)_{1/2,h}$ and we solve,

$$\begin{cases} a(u_h, v_h) + b(v_h, \lambda_h) = (f, v_h) & \forall v_h \in V_h, \\ b(u_h, \mu_h - \lambda_h) \geq (g_h, \mu_h - \lambda_h)_{1/2,h} & \forall \mu_h \in \Lambda_h. \end{cases}$$

If we denote $\mu_h^+ \geq 0$ the elements of Λ_h positive at the nodes and write $r_h = \mathcal{B}_h u_h - g_h$ the discrete Kuhn-Tucker conditions are

$$\begin{cases} \lambda_h^+ \in \Lambda_h^+ \\ (r_h, \mu_h^+)_{1/2,h} \geq 0 & \forall \mu_h^+ \\ (r_h, \lambda_h^+)_{1/2,h} = 0 \end{cases}$$

Referring to Sect. 5.1.3, this can be written in terms of nodal values,

$$\langle Br, \mu^+ \rangle \geq 0 \quad \forall \mu^+ \qquad \langle Br, \lambda^+ \rangle = 0.$$

This means that r is positive in some average sense.

The Active Set Strategy

In Sect. 5.2 we shall us the *active set strategy* [3, 54, 55] which we already discussed in Sect. 3.1.3

This is an iterative procedure, determining the zone where the equality condition $u = g$ must be imposed. We define the **contact status** dividing Γ_C in two parts.

- Active zone : $\lambda_h > 0$ or $\lambda_h = 0$ and $Br_h < 0$
- Inactive zone : $\lambda_h = 0$ and $Br_h \geq 0$

Remark 5.10 (A Sequence of Unconstrained Problems) Once the contact status is
determined, one solves on the active zone **equality constraints** $Br_h = 0$ using
algorithms Algorithm 3.7 with Algorithm 3.10. One could also, as in Sect. 5.1.4,
take an approximate Steklov-Poincaré operator on the active zone in Algorithm 5.1.
We shall discuss this in more detail in Remark 5.13.

The active zone is checked during the iteration and if it changes, the iterative process
is reinitialised.
   We shall illustrate below this procedure for contact problems.

5.2 Sliding Contact

We consider, for example as in Fig. 5.3, an elastic body in contact with a rigid
surface. We shall restrict ourselves to the case of frictionless contact as the frictional
case would need a much more complex development. We refer to [34] for a more
general presentation. We thus look for a displacement $\underline{u}$ minimising some elasticity
potential $J(\underline{v})$ under suitable boundary conditions. In the case of linear elasticity,
we would have

$$J(\underline{v}) = \mu \int_\Omega |\underline{\underline{\varepsilon}}(\underline{v})|^2 \, dx - \int_\Omega \underline{f} \cdot \underline{v} \, dx.$$

   Here we take $\underline{v} \in V \subset (H^1(\Omega))^3$, the space of displacements. In the following
we denote,

- Γ_C is a part of the boundary where contact is possible.
- $\mathcal{B}$ is the trace operator of the elements of V in $\Lambda \subset (H^{1/2}(\Gamma_C))^3$.

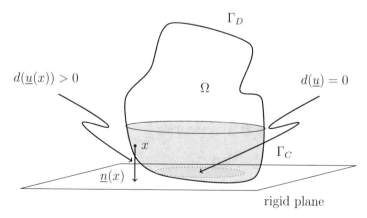

Fig. 5.3 Contact with a horizontal rigid plane. An elastic body Ω submitted to a displacement $\underline{u}$.
Illustration of the oriented distance computed on the potential contact surface Γ_C

- $\mathcal{R}$ is the Ritz operator from Λ onto Λ'.
- $B = \mathcal{R}\mathcal{B}$ is an operator from V into Λ'.

The basic quantity for the contact problem is the oriented distance, computed by projecting the elastic body on a target surface. The oriented distance to this target is defined as negative in case of penetration and this projection implicitly define the normal vector $\underline{n}$. Since $\underline{u}$ is a displacement of the body Ω, the oriented distance is in fact a function of $\underline{u}$, we will express this dependency by denoting it $d(\underline{u})$.

The distance $d(\underline{u})$ is a function on the boundary which belongs to a functional space which we assume to be a subspace Λ of $H^{1/2}(\Gamma_C)$. If the boundary is not smooth (e.g. Lipschitzian), we have to introduce a space defined on each smooth part. We place ourselves in the setting of Sect. 2.1.5. In this context $\mathcal{B}\underline{v}$ would be $d(\underline{v})$ but in general $d(\underline{v})$ is non linear so that we will need a linearisation to apply our algorithms. The first (non linear) condition to be respected is then the non penetration, that is

$$d(\underline{u}) \geq 0 \ on \ \Gamma_C$$

and we can write the first step of our model as,

$$\inf_{d(\underline{v}) \geq 0} J(\underline{v}).$$

Contact Pressure
The next step is to introduce a Lagrange multiplier $\underline{\lambda} \in \Lambda$ for the constraint: the contact pressure [35, 59]. We thus transform our minimisation problem into the saddle-point problem,

$$\inf_{\underline{v}} \sup_{\lambda_n \geq 0} J(\underline{v}) + (\lambda_n, d(\underline{v}))_\Lambda.$$

The optimality conditions are then

$$\begin{cases} \langle A(\underline{u})\,\underline{u} - \underline{f}, \underline{v} \rangle + (\lambda_n, \underline{v} \cdot \underline{n})_\Lambda = 0 & \forall \underline{v} \\ (d(\underline{u}), \mu_n - \lambda_n)_\Lambda \geq 0 & \forall \mu_n \geq 0. \end{cases}$$

where the operator $A(\underline{u})$ represent the constitutive law of the material. From this system we deduce the Kuhn-Tucker conditions,

$$\begin{cases} \lambda_n \geq 0, \\ (d(\underline{u}), \mu)_\Lambda \geq 0 & \forall \mu \geq 0, \\ (d(\underline{u}), \lambda_n)_\Lambda = 0. \end{cases}$$

Newton's Method, Sequential Quadratic Programming

It must be noted that even in the case of linear elasticity, the problem is non linear. To solve this problem, we apply the Newton method or in this context the *sequential quadratic programming* (SQP) method [58]. Let $\underline{u}^0$ be some initial value of $\underline{u}$ and $g_n^0 = d(\underline{u}^0)$ the corresponding initial gap. We recall [33] that the derivative of the distance function is given by

$$d'(\underline{u}_0) \cdot \delta\underline{u} = \delta\underline{u} \cdot \underline{n}$$

The linearised problem is then,

$$
\begin{cases}
\langle A'(\underline{u}^0)\delta\underline{u}, \underline{v}\rangle + (\lambda_n, \underline{v} \cdot \underline{n})_\Lambda = \langle \underline{f} - A(\underline{u}^0), \underline{v}\rangle & \forall \underline{v} \in V \\
(\phi_n - \lambda_n, g_n^0 - \delta\underline{u} \cdot \underline{n})_\Lambda \geq 0 & \forall \phi_n \geq 0
\end{cases}
\tag{5.17}
$$

Here $\delta\underline{u}$ is the correction of the initial value $\underline{u}^0$, and we have linearised both the constitutive law of the material represented by the operator A and the distance function. The Kuhn Tucker conditions then become

$$
\begin{cases}
\lambda_n \geq 0, \\
(g_n^0 - \delta\underline{u} \cdot \underline{n}, \mu_n)_\Lambda \geq 0 & \forall \mu_n \geq 0, \\
(g_n^0 - \delta\underline{u} \cdot \underline{n}, \lambda_n)_\Lambda = 0.
\end{cases}
$$

Remark 5.11 (The Choice of Multipliers)   In the above formulation, we have worked with the bilinear form

$$b(\underline{v}, \lambda_n) = (\underline{v} \cdot \underline{n}, \lambda_n)_\Lambda$$

that is with the scalar product in $\Lambda \subset H^{1/2}(\Gamma_C)$. As we have seen in Remark 5.3 we can also write (5.17) with

$$\langle \lambda', \underline{v} \cdot \underline{n}\rangle = \langle \mathcal{R}\lambda, \underline{v} \cdot \underline{n}\rangle$$

We recall that this is the same formulation written in two different ways.   ∎

5.2.1   The Discrete Contact Problem

In the previous Sect. 5.2 we developed a formulation of sliding contact using a Lagrange multiplier. We have presented a simpler problem in Sect. 5.1. We now come to the contact problem. We suppose that we have chosen a space V_h to discretise displacements and a space $\Lambda_h \subset \Lambda$. The choice of these spaces must of course be done to satisfy some requirements.

- The material will often be incompressible : we shall consider this case in Chap. 6. This imposes some restrictions on the choice of V_h and of the associated pressure space Q_h. We refer to [20] for details. In practice, for three-dimensional problems, we use as in Sect. 4.2 the Taylor-Hood approximation, that is piecewise quadratic elements for V_h and piecewise linear for Q_h, but this is one of many possibilities.
- The problem (5.17) is a (constrained) saddle-point problem. This means that the choice of Λ_h requires an inf-sup condition to be satisfied, [13]. In the following, we use piecewise linear elements for Λ_h.

Remark 5.12 (Scalar Product on Λ_h)   We also have to define a scalar product on Λ_h. Ideally this should be the scalar product in Λ. As we have discussed earlier the scalar product in $H^{1/2}$ is not easily computed and we shall have to employ some approximation. In general we consider as in Section 5.1.5 a discrete $H^{1/2}$ norm defined by a discrete scalar product $(\lambda_h, \mu_h)_{1/2,h}$.

We thus have an operator $\mathcal{R}_h$ from Λ_h onto Λ'_h

$$\langle \mathcal{R}_h \lambda_h, \mu_h \rangle_{\Lambda'_h \times \Lambda_h} = (\lambda_h, \mu_h)_{1/2 \cdot h}.$$

and the associated matrix R. In our computations we shall rely on Algorithm 3.10. An important issue is the choice of M_S and its inverse M_S^{-1}. the normal choice is here $M_S = R$. From Proposition 2.2, to obtain a convergence independent of the mesh size, R should define on Λ_h a scalar product coherent with the scalar product of $H^{1/2}$ in order to have an inf-sup condition independent of h.

We thus have to make a compromise: a better R yields a better convergence but may be harder to compute.

In the results presented below, we use the simple approximation by the L^2 scalar product in Λ_h and the matrix R becomes M_0 defined by,

$$\langle M_0 \lambda, \mu \rangle = \int_C \lambda_h \, \mu_h ds. \tag{5.18}$$

We could also employ a diagonal approximation M_D to M_0. This can be obtained by replacing (5.18) by a Newton-Cotes quadrature using only the nodes of Λ_h. In the case of piecewise linear elements, this is equivalent to using a 'lumped'(diagonal) matrix obtained by summing the rows of M.   ∎

The geometrical nodes of the mesh which support degrees of freedom for λ_h and u are not identical. To emphasise this difference we denote by $\underline{x}_i$ the nodes supporting the values of $\underline{\lambda}_h \in \Lambda_h$ and $\underline{y}_j$ the nodes for $\underline{u}_h$. As in Remark 2.5, we now denote $\underline{u}$ and $\underline{\lambda}$ the vectors of nodal values, whenever no ambiguity arises. We thus have

$$\underline{u} = \{\underline{u}_h(\underline{y}_j), 1 \leq j \leq N_V\}$$

$$\underline{\lambda} = \{\underline{\lambda}_h(\underline{x}_i), 1 \leq i \leq N_\Lambda\}$$

Denoting by $\langle \cdot, \cdot \rangle$ the scalar product in $\mathbb{R}^{N_V}$ or $\mathbb{R}^{N_\Lambda}$, we thus have the matrices associated with operators $\mathcal{R}_h$ and $\mathcal{B}_h$

$$\langle R\underline{\lambda}, \underline{\mu} \rangle = (\underline{\lambda}_h, \underline{\mu}_h)_{1/2,h}, \quad R \in \mathbb{R}^{N_\Lambda \times N_\Lambda}$$

$$\langle B\underline{u}, \underline{\mu} \rangle = (\mathcal{B}_h\underline{u}_h, \underline{\mu}_h)_{1/2,h}, \quad B \in \mathbb{R}^{N_\Lambda \times N_V}$$

and then $R^{-1}B$ is the matrix associated with the projection of $\mathcal{B}u_h$ on Λ_h.

• These matrices depend on the choice of the discrete scalar product.

We also suppose that we have a normal $\underline{n}_i$ defined at $\underline{x}_i$ and we define for $\underline{\lambda}_h$ its normal component such that at all nodes,

$$\lambda_{nh}(\underline{x}_i) = \underline{\lambda}_h(\underline{x}_i) \cdot \underline{n}_i$$

which is a scalar. We also use its vectorial version

$$\underline{\lambda}_{nh}(\underline{x}_i) = \lambda_n(\underline{x}_i)\underline{n}_i \tag{5.19}$$

We denote by Λ_{nh} the subset of Λ_h of normal vectors of the form (5.19) and

$$\Lambda_{nh}^+ = \{\underline{\lambda}_{nh} \in \Lambda_{nh} \mid \lambda_n(\underline{x}_i) \geq 0\}$$

We want to have as in Sect. 5.1.7 a discrete form of the condition $\underline{u} \cdot \underline{n} \geq 0$. Recall that $B\underline{u}$ is a vector corresponding to an element of Λ'_h. We consider at node $\underline{x}_i$ its normal component

$$B_n\underline{u} = \{(B\underline{u})_i \cdot \underline{n}_i\}$$

We then have the condition

$$\langle B_n\underline{u}, \underline{\mu}_n^+ \rangle \geq 0 \quad \forall \underline{\mu}_n^+.$$

which is a weak form of $\underline{u} \cdot \underline{n} \geq 0$.

We can now introduce our discrete contact problem. We first consider the unconstrained case, $B_n\underline{u} = g$ where we do not have to consider a positivity condition. Indeed our final algorithm will solve a sequence of such unconstrained problems, we thus look for the solution of

$$\inf_{\underline{v}} \sup_{\underline{\lambda}_n} \frac{1}{2} \langle A\underline{v}, \underline{v} \rangle + \langle R\underline{\lambda}_n, (R^{-1}B\underline{v} - \underline{g}) \rangle - \langle \underline{f}, \underline{v} \rangle \qquad (5.20)$$

that is also

$$\inf_{\underline{v}} \sup_{\underline{\lambda}_n} \frac{1}{2} \langle A\underline{v}, \underline{v} \rangle + \langle \lambda_n, (B_n \underline{v} - R\underline{g}) \rangle - \langle \underline{f}, \underline{v} \rangle. \qquad (5.21)$$

This problem can clearly be solved by the algorithms of Chap. 3. We must however introduce a way to handle the inequality constraint. To do this we first need the notion of contact status.

Contact Status

Let $r_n = B_n \underline{u} - g$ the normal residual. A basic tool in the algorithms which follow, will be the contact status. It will be represented point wise by the operator $P(\underline{\lambda}, r_n)$ defined by,

$$\begin{cases} \text{if } \lambda_n = 0, \\ \qquad (1) \text{ if } r_n \leq 0 \text{ then } P(\underline{\lambda}, r_n) = 0 \\ \qquad (2) \text{ if } r_n > 0 \text{ then } P(\underline{\lambda}, r_n) = r_n \\ \text{if } \lambda_n > 0, \\ \qquad (3) \ P(\underline{\lambda}, r_n) = r_n \end{cases}$$

The Kuhn-Tucker optimality conditions can then written as $P(\underline{\lambda}, r_n) = 0$. We say that in case (1), the constraints are inactive and that on (2) and (3), they are active. We denote Γ_{C_A} subset of Γ_C where the constraints are active.

5.2.2   The Algorithm for Sliding Contact

We can now present our solution strategy for the sliding contact problem.

A Newton Method

We have a nonlinear problem. As we have seen in Sect. 5.2, we rely on a Newton method in which the distance is linearised.

The Active Set Strategy

The algorithm proposed here is based on the *active set strategy*

- Given $(\underline{u}^0, \lambda_n^0)$, compute $P(\underline{r}_n^0, \lambda_n^0)$, that is the contact status.
- Solve (on the active set) an *equality constraint* problem.

$$\begin{cases} \langle A'(\underline{u}^0)\delta\underline{u}, \underline{v}\rangle + \langle \lambda_n, \underline{v}\rangle = \langle \underline{f}, \underline{v}\rangle & \forall \underline{v} \\ \langle \delta\underline{u} \cdot \underline{n} - g_n^0, \phi_n\rangle = 0 & \forall \phi_n \end{cases} \tag{5.22}$$

- Check the contact status after a few iterations.
  - If the status has changed: project λ on the admissible set Λ_h+; recompute the active and inactive sets; restart.
  - If the contact status does not change, we continue the iteration.
- Iterate till those sets are stabilised.

The contact status is thus checked during the iterations to avoid continuing to solve a problem which is no longer relevant.

Remark 5.13 (M_s and the Active Zone)   The active set strategy solves a sequence of problems with equality constraints on a **subset** Γ_{C_A} of Γ_C where the constraints are actually imposed. This will mean in Algorithm 3.10 to restrict M_S to Γ_{C_A}.

  Let us first consider the case where we would take $M_S = I$. One would then have a correction z_λ on λ_n given by $z_\lambda = r_n$. If we do not want to modify λ_n on the inactive zone we should make $r_n = 0$ on inactive nodes. This is analogue to the classical conjugate projected gradient method [69]. The same is true if one has $M_S = M_D$ where M_D is the diagonal matrix introduced in Remark 5.12.

  In general we can proceed by,

- Make $r_n = 0$ outside the active zone Γ_{C_A}. Solve $M_S z = r_n$ imposing $z = 0$ outside of Γ_{C_A}.

  If one uses the diagonal matrix M_D, this is automatically taken into account to the price of a slower convergence.

  A similar procedure should be used if one introduces a discrete Steklov-Poincaré operator.   ∎

5.2.3 *A Numerical Example of Contact Problem*

To illustrate the behaviour of the algorithm we present the case of an elastic cube being forced on a rigid sphere (see Fig. 5.4). In this case a displacement is imposed on the upper surface of the cube. We want to illustrate the behaviour of the algorithm as the mesh gets refined (see Table 5.1).

Fig. 5.4 Contact pressure
over the potential contact area

Table 5.1 Iteration's number and CPU time according to the total number of degrees of freedom
(dof) of u and the solving method for the primal system

	14,739 dof		107,811 dof		823,875 dof	
	# it.	CPU (s)	# it.	CPU (s)	# it.	CPU (s)
LU	32	6.85	39	208.46	41	13099
GCR(10,**HP-LU**)	40	8.96	42	59.06	53	477.36

We consider a linear elastic material with a Poisson coefficient $v = 0.3$ and
an elasticity modulus $E = 10^2$. The primal problem will be solved by either a
direct or an iterative method and we present at Table 5.1 the two usual performance
indicators i.s.e the total number of iterations and the computational time in seconds.
Here we take $R = M_0$ (Remark 5.12) for which we do not expect convergence to
be independent of h. There is room for improvement but even then the results show
that the method is suitable for large problems.

In Table 5.1 the iterations numbers show that even when using the L^2 scalar
product for $b(\underline{v}_h, \lambda_h)$, convergence of the dual problem is only slightly dependent of
the discretization. The large differences in computation time illustrate the important
gain in efficiency of the iterative solver for large problems.

It is also interesting to consider convergence curves of the normal residual and
the importance of the solver in $\underline{u}$. In Fig. 5.5 one compares the convergence for three
mesh sizes. Red dots indicate a change in contact status in the active set strategy.
The changes occur at the beginning of the process and the iteration is then restarted
on the new active region.. Most of the computational efforts consist in solving a

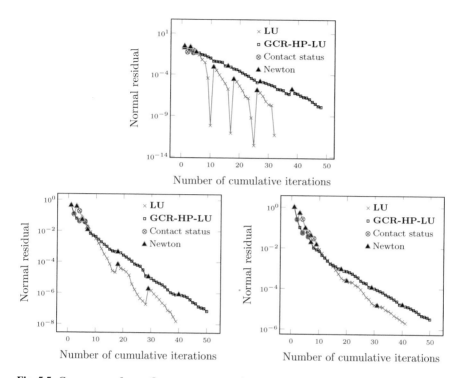

Fig. 5.5 Convergence for r_n. On top a coarse mesh (14,739 dof for u). Bottom left a mesh with 107,811 dof, on the right a mesh with 823,875 dof

single equality constraint problem at each Newton step. If we ignore computational time obviously the direct solver in $\underline{u}$ perform better since it produces less iterations compared to the iterative solver. Finally, for smaller mesh (top of Fig. 5.5) the direct solver seems too good: time is lost with excessive tolerance at each Newton step.

Chapter 6
Solving Problems with More Than One Constraint

In this chapter, we shall consider problems of the form

$$
\underbrace{\begin{pmatrix} A & B^t & C^t \\ B & 0 & 0 \\ C & 0 & 0 \end{pmatrix}}_{\mathcal{A}} \begin{pmatrix} u \\ p \\ \lambda \end{pmatrix} = \begin{pmatrix} f \\ g_p \\ g_\lambda \end{pmatrix}.
\tag{6.1}
$$

In theory, this is not fundamentally different from the problem (2.22). From the numerical point of view, things are not so simple as the presence of two independent constraints brings new difficulties in the building of algorithms and preconditioners. We present some ideas which indeed yield more questions than answers.

6.1 A Model Problem

To fix ideas, we consider a case where two simultaneous constraints are to be satisfied. We have already considered in Sect. 4.2 problems of incompressible elasticity and in Sect. 5 problems of sliding contact. We now want to solve problems where both these features are involved.

Let Ω be a domain of $\mathbb{R}^3$, Γ its boundary and Γ_C a part of Γ where sliding contact conditions are imposed by Lagrange multipliers. We suppose that suitable Dirichlet conditions are imposed on a part of the boundary to make the problem well posed.

We shall use the same procedure as in Sect. 5.2, the only difference will be that we now denote C the operator which was then denoted B, this operator now being associated with the discrete divergence: we now want the material to be incompressible. Proceeding again to linearise the problem, and using the active set

© The Author(s), under exclusive license to Springer Nature Switzerland AG 2022
J. Deteix et al., *Numerical Methods for Mixed Finite Element Problems*,
Lecture Notes in Mathematics 2318, https://doi.org/10.1007/978-3-031-12616-1_6

technique, we are led to solve a sequence of saddle-point problem of the form

$$\inf_{\underline{v}} \sup_{p, \lambda_n} \frac{1}{2} \langle A\underline{v}, \underline{v} \rangle + \langle \lambda_n, (C\underline{v} - r_\lambda) \rangle + \langle p, (B\underline{v} - r_p) \rangle - \langle \underline{r}_u, \underline{v} \rangle.$$

This is indeed a problem similar to (6.1). We shall first consider a naive solution technique, the interlaced method and then reconsider the use of approximate factorisations.

6.2   Interlaced Method

It would be natural to rewrite the system (6.1) as a problem with a single constraint. To do so we introduce block matrices

$$\begin{pmatrix} \mathfrak{A} & \mathfrak{C}^t \\ \mathfrak{C} & 0 \end{pmatrix} \begin{pmatrix} \mathfrak{u} \\ \underline{\lambda} \end{pmatrix} = \begin{pmatrix} \mathfrak{r}_u \\ \mathfrak{r}_\lambda \end{pmatrix}$$

where the blocks are defined by

$$\mathfrak{A} = \begin{pmatrix} A & B^t \\ B & 0 \end{pmatrix}, \quad \mathfrak{C} = \begin{pmatrix} C & 0 \end{pmatrix}, \quad \mathfrak{u} = \begin{pmatrix} u \\ p \end{pmatrix}, \quad \mathfrak{r}_u = \begin{pmatrix} r_u \\ r_p \end{pmatrix} \text{ and } \mathfrak{r}_\lambda = \begin{pmatrix} r_\lambda \\ 0 \end{pmatrix}.$$

It is then possible to handle this problem as a contact problem where each solution of the elasticity problem is in fact the solution of an incompressible elasticity problem which can be solved (with more or less precision) by the methods presented in Sect. 4.2. We call this the interlaced method.

Remark 6.1 The choice of the 'inner' problem, here the incompressible elasticity problem is arbitrary. We could also have chosen the contact problem. Our choice is directed by the fact that the incompressible elasticity problem is linear.   ∎

Starting from this rewriting of the system, we can use directly the technique of Sect. 3.2.1, in particular the preconditioner of Algorithm 3.10 in which we replace the approximate solver for A by an approximate solver for $\mathfrak{A}$. This is simple but carries some difficulties.

- The matrix $\mathfrak{A}$ is not definite positive. This essentially renders invalid the classical proofs of convergence, for example of Uzawa's method.
- We have to solve an internal problem which will make the method expensive.

The downside of this approach is thus the significant cost of the iterations in λ. In fact, at each iteration, two solutions to the problem in $(\underline{u}, p)$ are asked. These latter being made iteratively they also require at least two displacement only solutions.

Remark 6.2 (Projected Gradient) If we refer ourselves to Sect. 3.1.3 one sees that what we are doing is iterating in the divergence-free subspace provided the problems in $(\underline{u}, p)$ are solved precisely. ■

To illustrate the behaviour of the algorithm, we first consider the case of an accurate solution in $(\underline{u}, p)$. This will be done as in Sect. 4.2.4. For the solve in $\underline{u}$, we take either a direct LU solver or some iterations of GCR preconditioned by HP-IU. This is clearly an expensive procedure only used for comprehension.

We present the convergence in λ for the intermediate mesh of Sect. 5.2.3 with 107811 degrees of freedom.

As can be expected the results are comparable for the two solver in $\underline{u}$ as we use essentially the same information to update λ. Indeed if we have an accurate solution in $(\underline{u}, p)$ it should not be dependent on the way it is computed (Fig. 6.1).

When an incomplete solution in $(\underline{u}, p)$ is considered by limiting the number of iterations permitted in the Mixed-GMP-GCR method, the iterative solution in $\underline{\lambda}_n$ is loosing effectiveness (Fig. 6.2).

Fig. 6.1 Interlaced method using complete solver for $\mathfrak{A}$: convergence for the normal residual according to the primal solver with the intermediate mesh

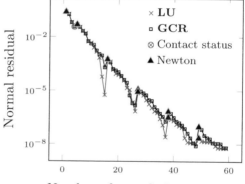

Fig. 6.2 Interlaced method using incomplete solver for $\mathfrak{A}$: convergence according to the primal solving method using intermediate mesh

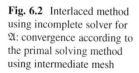
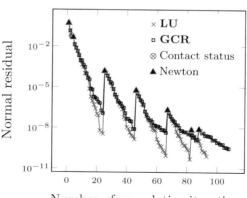

		107 811 dof
		#i$CPU(s)$
LU	93	464.64
GCR(10,**HP**)	108	4928.12

Table 6.1 Interlaced method using incomplete solver for $\mathfrak{A}$: number of iterations and CPU time in seconds according to the solver in $\underline{u}$

Although it works, it is clearly an expensive method (Table 6.1) and we did not push the test to finer meshes. We now consider another approach which looked more promising.

6.3   Preconditioners Based on Factorisation

In Sect. 3.2.1 we presented a general preconditioner based on the factorisation of the matrix associated to our problem. This can be done in the present case but will also imply to develop variants. It is easily seen that the matrix of equation (6.1) can be factorised in the form,

$$
\mathcal{A} = \begin{bmatrix} I & 0 & 0 \\ BA^{-1} & I & 0 \\ CA^{-1} & 0 & I \end{bmatrix} \begin{bmatrix} A & 0 & 0 \\ 0 & -S_{BB} & -S_{CB} \\ 0 & -S_{CB} & -S_{CC} \end{bmatrix} \begin{bmatrix} I & A^{-1}B^t & A^{-1}C^t \\ 0 & I & 0 \\ 0 & 0 & I \end{bmatrix}
$$

where

$$
S_{BB} = BA^{-1}B^t, \quad S_{BC} = BA^{-1}C^t, \quad S_{CB} = CA^{-1}B^t, \quad S_{CC} = BA^{-1}C^t.
$$

We shall denote

$$
S = \begin{pmatrix} S_{BB} & S_{BC} \\ S_{CB} & S_{CC} \end{pmatrix}
$$

We now see what the Algorithm 3.7 would now become,

Algorithm 6.1 *General mixed preconditioner for two constraints Assuming that we have $\widetilde{A}^{-1}$ and $\widetilde{S}^{-1}$ approximate solvers for A and S, from the residual r_u, r_p and r_λ, we compute the vectors z_u, z_p and z_λ.*

- *Initialise*

$$
z_u^* = \widetilde{A}^{-1} r_u
$$
$$
r_p = B z_u^* - r_p,
$$
$$
r_\lambda - C z_u^* - r_\lambda. \tag{6.1}
$$

- *Using $\widetilde{\mathcal{S}}^{-1}$, solve approximately for z_p and z_λ the coupled problem,*

$$\begin{pmatrix} S_{BB} & S_{BC} \\ S_{CB} & S_{CC} \end{pmatrix} \begin{pmatrix} z_p \\ z_\lambda \end{pmatrix} = \begin{pmatrix} r_p \\ r_\lambda \end{pmatrix} \tag{6.2}$$

- *Compute*

$$z_u = z_u^* - \widetilde{A}^{-1} B^t z_p - \widetilde{A}^{-1} C^t z_\lambda \tag{6.3}$$

∎

The key is thus to solve (6.2). There is clearly no general way of doing this. Starting from the point of view that we have approximate solvers for S_{BB} and S_{CC}, we may use a Gauss-Seidel iteration for z_p and z_λ.

In the simplest case we do one iteration for each problem.

6.3.1   The Sequential Method

In this case we use

- One iteration in p and one in λ
- The result is used in a GCR global method.

In the following Table 6.2, we present the number of iterations and the solution time, for different ways of solver for the matrix A of the primal problem.

For small meshes, LU factorisation is clearly the best way and this also the true in all cases with respect to iteration number. A GCR method without preconditioning rapidly deteriorates. The multigrid HP-AMG method becomes the best for computing time for the fine mesh.

Looking to convergence curves, we see that even this last method could be improved as the convergence seems to slow down when the residual becomes small. For a still finer mesh one could need a more accurate solver for A to avoid

Table 6.2 Number of iterations and CPU time (s) according to the size of the mesh and the solving method for the primal system

	14,739 dof		107,811 dof		823,875 dof	
	#it.	CPU (s)	#it.	CPU (s)	# it.	CPU (s)
LU	55	8.36	88	120.94	124	3432.87
GCR(10)	129	92.97	141	724.54	209	7940.04
CG(10,GAMG)	129	62.98	99	332.34	162	4459.59
GCR(5,HP-AMG)	174	36.66	159	191.78	159	1533.71

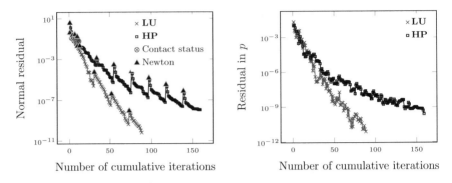

Fig. 6.3 Convergence curves for the normal residual (left) and hydrostatic pressure p (right) according to the primal problem solving method when the middle mesh is considered

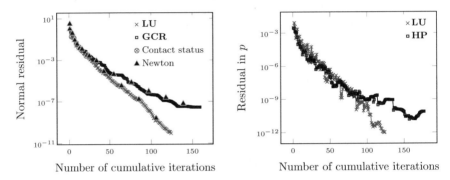

Fig. 6.4 Convergence curves for the normal residual (left) and hydrostatic pressure p (right) according to the primal problem solving method when the finest mesh is considered

stagnation, keeping in mind that LU becomes more and more inefficient (Figs. 6.3 and 6.4).

6.4 An Alternating Procedure

Finally, to end this presentation, we present some results for an 'Alternating Direction' procedure. The idea is simple: solve alternatingly the contact problem for $(\underline{u}, \lambda_n)$ fixing the hydrostatic pressure p and then solve for $(\underline{u}, p)$ fixing the contact pressure λ_n. Another issue is that for the contact part, a restart procedure has to be considered when a change occurs in contact status (Fig. 6.5).

Fig. 6.5 Illustration of the alternating procedure

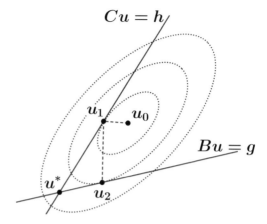

Table 6.3 Alternating method: iteration's number/CPU(s) according to the size of the mesh and the solving method for the primal system

	14,739 dof		107,811 dof		823,875 dof	
	# it.	CPU (s)	# it.	CPU (s)	# it.	CPU (s)
LU	66	7	110	93	120	2585
GCR(10)	110	35	120	294	150	2966
GCR(5,HP-LU)	120	17	150	139	160	1168

Remark 6.3 (A Projection Method) This could be seen as a Projection method. After a few iterations in Λ, the solution is projected on the divergence-free subspace. ∎

As we did in the previous section with the sequential method, some computations are presented with this alternating method for the three different meshes. In the Table 6.3, we present the number of global iteration on λ and the computing time for different solvers for the primal problem. The conclusion is the same as for all previous examples: when the mesh size gets large, the iterative method (in particular when using the HP-LU method of Sect. 3.1.4) becomes better in computing time.

For the convergence, we can see in Figs. 6.6 and 6.7 respectively the middle and finest mesh, the curves for the hydrostatic and contact pressure according to the solver of primal problem. We see a good convergence of thoses curves particulary the contact one.

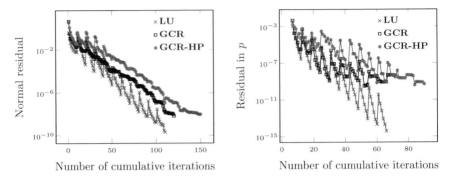

Fig. 6.6 Alternating method: convergence curves for the normal residual (left) and hydrostatic pressure (p) according to the primal problem solving method when the middle mesh is considered

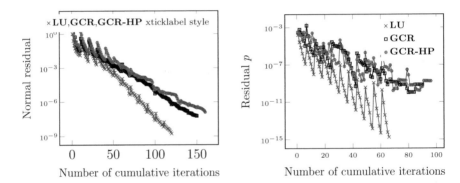

Fig. 6.7 Alternating method: convergence curves for the normal residual (left) and p the hydrostatic pressure (right) according to the primal problem solving method when the finest mesh is considered

This is a crude first testing and things could clearly be ameliorated. In particular, one could think of marrying the alternating idea into the sequential technique of the previous Sect. 6.3.1. There is room for new ideas...

Chapter 7
Conclusion

We hope to have shown that solving mixed problems can be accomplished efficiently. This work is clearly not exhaustive and we have indeed tried to open the way for future research. We have relied as building bricks on rather classical iterative methods. However, we think that we have assembled these bricks in some new ways. We have also insisted in developing methods as free as possible of user depending parameters.

We have also considered Augmented Lagrangians, either in an exact or a regularised version. This was done in the mind that direct methods should be avoided for large scale computations and that penalty terms destroy the condition number of the penalised system.

- For mixed formulations based on elements satisfying the equilibrium conditions, we have shown that Augmented Lagrangian is efficient. The problem presented was very simple but we think that the results could be extended to more realistic situations. In the case of mixed elasticity in which a symmetry condition has to be imposed [19] one would have to deal with two constraints as in Chap. 6. The situation would be better than in the example presented there as the equilibrium constraint is amenable to a real augmented Lagrangian.
- Problems involving incompressible elasticity are of central importance in many applications. Unfortunately, they are often solved with poor methods using penalty and low order elements. We have shown that continuous pressure elements, which are essential for accurate three-dimensional computations at reasonable cost, are manageable and can be accelerated by a stabilisation term.
- For contact problems, we have considered some possible alternative avenues to the standard approximations where the constraint is treated in L^2 instead of the correct $H^{1/2}$. This is still an open area. We have shown that using the more classical formulation, one can obtain results for large meshes with good efficiency.
- For problems involving two constraints, we have explored some possibilities and many variants are possible.

© The Author(s), under exclusive license to Springer Nature Switzerland AG 2022
J. Deteix et al., *Numerical Methods for Mixed Finite Element Problems*,
Lecture Notes in Mathematics 2318, https://doi.org/10.1007/978-3-031-12616-1_7

The methods that we discussed can in most cases be employed for parallel computing. We did not adventure ourselves in this direction which would need a research work by itself. We must nevertheless emphasise that the Petsc package which we used is intrinsically built for parallel computing.

Bibliography

1. B. Achchab, J.F. Maître, Estimate of the constant in two strengthened C.B.S. inequalities for F.E.M. systems of 2D slasticity: application to multilevel methods and a posteriori error estimators. Numer. Linear Algebra Appl. **3**(2), 147–159 (1996)
2. ADINA Inc., Instability of two-term mooney-rivlin model. https://www.adina.com/newsgH48.shtml
3. P. Alart, A. Curnier, A mixed formulation for frictional contact problems prone to Newton like solution methods. Comput. Methods Appl. Mech. Eng. **92**(3), 353–375 (1991)
4. G. Allaire, S.M. Kaber, *Numerical Linear Algebra*. Texts in Applied Mathematics, vol. 55 (Springer, New York, 2008)
5. M. Arioli, D. Kourounis, D. Loghin, Discrete fractional Sobolev norms for domain decomposition preconditioning. IMA J. Numer. Anal. **33**(1), 318–342 (2013)
6. D.N. Arnold, F. Brezzi, Mixed and nonconforming finite element methods: implementation, postprocessing and error estimates. ESAIM: Math. Model. Numer. Anal. **19**(1), 7–32 (1985)
7. W.E. Arnoldi, The principle of minimized iterations in the solution of the matrix eigenvalue problem. Quart. Appl. Math. **9**(1):17–29 (1951)
8. K.J. Arrow, L. Hurwicz, H. Uzawa, *Studies in Linear and Non-linear Programming*. Stanford Mathematical Studies in the Social Sciences, vol. 2 (Stanford University Press, Stanford, 1958)
9. O. Axelsson, G. Lindskog, On the rate of convergence of the preconditioned conjugate gradient method. Numer. Math. **48**(5), 499–523 (1986)
10. Z.Z. Bai, B.N. Parlett, Z.Q. Wang, On generalized successive overrelaxation methods for augmented linear systems. Numer. Math. **102**(1), 1–38 (2005)
11. S. Balay, S. Abhyankar, M. Adams, J. Brown, P. Brune, K. Buschelman, L. Dalcin, A. Dener, V. Eijkhout, W. Gropp, D. Karpeyev, D. Kaushik, M. Knepley, D. May, L. Curfman McInnes, R. Mills, T. Munson, K. Rupp, P. Sanan, B. Smith, S. Zampini, H. Zhang, H. Zhang, *PETSc Users Manual: Revision 3.10*. (Argonne National Lab. (ANL), Argonne, 2018)
12. J. Baranger, J.F. Maitre, F. Oudin, Connection between finite volume and mixed finite element methods. ESAIM Math. Model. Numer. Anal. **30**(4), 445–465 (1996)
13. K.J. Bathe, F. Brezzi, Stability of finite element mixed interpolations for contact problems. Atti della Accademia Nazionale dei Lincei. Classe di Scienze Fisiche, Matematiche e Naturali. Rendiconti Lincei. Matematica e Applicazioni **12**(3), 167–183 (2001)
14. L. Beilina, E. Karchevskii, M. Karchevskii, Solving systems of linear equations, in *Numerical Linear Algebra: Theory and Applications* (Springer, Cham, 2017), pp. 249–289
15. M. Benzi, G.H. Golub, J. Liesen, Numerical solution of saddle point problems. Acta Numer. **14**, 1–137 (2005)

© The Author(s), under exclusive license to Springer Nature Switzerland AG 2022
J. Deteix et al., *Numerical Methods for Mixed Finite Element Problems*,
Lecture Notes in Mathematics 2318, https://doi.org/10.1007/978-3-031-12616-1

16. M. Bergounioux, K. Kunisch, Primal-dual strategy for state-constrained optimal control problems. Comput. Optim. Appl. **22**(2), 193–224 (2002)
17. M. Bergounioux, M. Haddou, M. Hintermüller, K. Kunisch, A comparison of a Moreau–Yosida based active set strategy and interior point methods for constrained optimal control problems. SIAM J. Optim. **11**(2), 495–521 (2000)
18. D.P. Bertsekas, *Nonlinear Programming*, 2nd edn. (Athena Scientific, Belmont, 1999)
19. D. Boffi, F. Brezzi, M. Fortin, Reduced symmetry elements in linear elasticity. Commun. Pure Appl. Anal. **8**(1), 95–121 (2009)
20. D. Boffi, F. Brezzi, M. Fortin, *Mixed Finite Element Methods and Applications*. Springer Series in Computational Mathematics, vol. 44 (Springer, Berlin 2013)
21. J. Bonet, R.D. Wood, *Nonlinear Continuum Mechanics for Finite Element Analysis*, 2nd edn. (Cambridge University Press, Cambridge, 2008)
22. D. Braess, Finite elements: theory, fast solvers and applications in solid mechanics. Meas. Sci. Technol. **13**(9), 365 (2007)
23. J.H. Bramble, J.E. Pasciak, A.T. Vassilev, Analysis of the inexact Uzawa algorithm for saddle point problems. SIAM J. Numer. Anal. **34**(3), 1072–1092 (1997)
24. F. Brezzi, K.J. Bathe, A discourse on the stability conditions for mixed finite element formulations. Comput. Methods Appl. Mech. Eng. **82**(1), 27–57 (1990)
25. F. Brezzi, M. Fortin, L.D. Marini, Mixed finite element methods with continuous stresses. Math. Models Methods Appl. Sci. **03**(02), 275–287 (1993)
26. J. Cahouet, J.P. Chabard, Some fast 3D finite element solvers for the generalized Stokes problem. Int. J. Numer. Methods Fluids **8**, 869–895 (1988)
27. M. Chanaud, Conception d'un solveur haute performance de systèmes linéaires creux couplant des méthodes multigrilles et directes pour la résolution des équations de Maxwell 3D en régime harmonique discrétisées par éléments finis. Ph.D. Thesis, Université de Bordeaux 1, France (2011). https://www.theses.fr/2011BOR14324/abes
28. S.H. Cheng, Symmetric indefinite matrices: linear system solvers and modified inertia problems. Ph.D. Thesis, University of Manchester (1998). https://www.maths.manchester.ac.uk/~higham/links/theses/cheng98.pdf
29. A.J. Chorin, Numerical solution of the Navier–Stokes equations. Math. Comput. **22**(104), 745–762 (1968)
30. P.G. Ciarlet, *The Finite Element Method for Elliptic Problems*. Studies in Mathematics and Its Applications (North-Holland, Amsterdam, 1978)
31. G.E. Comi, G. Stefani, A distributional approach to fractional Sobolev spaces and fractional variation: existence of blow-up. J. Function. Anal. **277**(10), 3373–3435 (2019)
32. M. Dauge, *Elliptic Boundary Value Problems on Corner Domains*. Lecture Notes in Mathematics, vol. 1341 (Springer, Berlin, 1988)
33. M.C. Delfour, J.P. Zolésio, *Shapes and Geometries: Analysis, Differential Calculus and Optimization*. Advances in Design and Control (Society for Industrial and Applied Mathematics, Philadelphia, 2001)
34. T. Diop, Résolution itérative de problèmes de contact frottant de grande taille. Ph.D. Thesis, Université Laval, Canada (2019). https://corpus.ulaval.ca/jspui/handle/20.500.11794/34968
35. G. Duvaut, J.L. Lions, *Inequalities in Mechanics and Physics*. Grundlehren der mathematischen Wissenschaften, vol. 219 (Springer, Berlin, 1976)
36. S.C. Eisenstat, H.C. Elman, M.H. Schultz, Variational iterative methods for nonsymmetric systems of linear equations. SIAM J. Numer. Anal. **20**(2), 345–357 (1983)
37. A. El maliki, R. Guénette, M. Fortin, An efficient hierarchical preconditioner for quadratic discretizations of finite element problems. Numer. Linear Algebra Appl. **18**(5), 789–803 (2011)
38. A. El maliki, M. Fortin, J. Deteix, A. Fortin, Preconditioned iteration for saddle-point systems with bound constraints arising in contact problems. Comput. Methods Appl. Mech. Eng. **254**, 114–125 (2013)
39. H.C. Elman, Iterative methods for large, sparse, nonsymmetric systems of linear equations. Ph.D. Thesis, Yale University (1982). http://www.cvc.yale.edu/publications/techreports/tr229.pdf.

40. H.C. Elman, G.H. Golub, Inexact and preconditioned Uzawa algorithms for saddle point problems. SIAM J. Numer. Anal. **31**(6), 1645–1661 (1994)
41. H.C. Elman, D.J. Silvester, A.J. Wathen, *Finite Elements and Fast Iterative Solvers: With Applications in Incompressible Fluid Dynamics*. Numerical Mathematics and Scientific Computation (Oxford University Press, Oxford, 2005)
42. A. Fortin, A. Garon, Les Élements Finis de La Théorie à La Pratique. GIREF, Université Laval (2018)
43. M. Fortin, R. Glowinski, *Augmented Lagrangian Methods: Applications to the Numerical Solution of Boundary-Value Problems*. Studies in Mathematics and its Applications, vol. 15 (North-Holland, Amsterdam, 1983)
44. M.J. Gander, Optimized Schwarz methods. SIAM J. Numer. Anal. **44**(2), 699–731 (2006)
45. A. Ghai, C. Lu, X. Jiao, A comparison of preconditioned Krylov subspace methods for large-scale nonsymmetric linear systems. Numer. Linear Algebra Appl. **26**(1), e2215 (2019)
46. G.H. Golub, C. Greif, On solving block-structured indefinite linear systems. SIAM J. Sci. Comput. **24**(Part 6), 2076–2092 (2003)
47. G.H. Golub, C.F. Van Loan, *Matrix Computations*, Johns Hopkins Studies in the Mathematical Sciences, 3rd edn. (Johns Hopkins University Press, Baltimore, 1996)
48. P.P. Grinevich, M.A. Olshanskii, An iterative method for the stokes-type problem with variable viscosity. SIAM J. Sci. Comput. **31**(5), 3959–3978 (2010)
49. A. Günnel, R. Herzog, E. Sachs, A note on preconditioners and scalar products in Krylov subspace methods for self-adjoint problems in Hilbert space. Electron. Trans. Numer. Anal. **41**, 13–20 (2014)
50. W. Hackbusch, *Iterative Solution of Large Sparse Systems of Equations*. Applied Mathematical Sciences, vol. 95, 2nd edn. (Springer, Amsterdam, 2016)
51. R. Herzog, K.M. Soodhalter, A modified implementation of Minres to monitor residual subvector norms for block systems. SIAM J. Sci. Comput. **39**(6), A2645–A2663 (2017)
52. M.R. Hestenes, E. Stiefel, Methods of conjugate gradients for solving linear systems. J. Res. Natl. Bureau Standards **49**(6), 409–436 (1952)
53. G.A. Holzapfel, *Nonlinear Solid Mechanics : A Continuum Approach for Engineering* (Wiley, New York, 2000)
54. S. Hüeber, B.I. Wohlmuth, A primal-dual active set strategy for non-linear multibody contact problems. Comput. Methods Appl. Mech. Eng. **194**(27), 3147–3166 (2005)
55. S. Hüeber, G. Stadler, B.I. Wohlmuth, A primal-dual active set algorithm for three-dimensional contact problems with coulomb friction. SIAM J. Sci. Comput. **30**(2), 572–596 (2009)
56. K. Ito, K. Kunisch, Augmented Lagrangian methods for nonsmooth, convex optimization in Hilbert spaces. Nonlinear Anal. **41**(5), 591–616 (2000)
57. K. Ito, K. Kunisch, Optimal control of elliptic variational inequalities. Appl. Math. Optim. Int. J. Appl. Stoch. **41**(3), 343–364 (2000)
58. E.G. Johnson, A.O. Nier, Angular aberrations in sector shaped electromagnetic lenses for focusing beams of charged particles. Phys. Rev. **91**(1), 10–17 (1953)
59. N. Kikuchi, J.T. Oden, *Contact Problems in Elasticity: A Study of Variational Inequalities and Finite Element Methods*. SIAM Studies in Applied Mathematics, vol. 8 (SIAM, Philadelphia, 1988)
60. C. Lanczos, An iteration method for the solution of the eigenvalue problem of linear differential and integral operators. J. Res. Natl. Bureau Standards **45**, 255–282 (1950)
61. S. Léger, Méthode lagrangienne actualisée pour des problèmes hyperélastiques en très grandes déformations. Ph.D. Thesis, Université Laval, Canada (2014). https://corpus.ulaval.ca/jspui/handle/20.500.11794/25402
62. J.L. Lions, E. Magenes, *Non-Homogeneous Boundary Value Problems and Applications*. Grundlehren der mathematischen wissenschaften; bd. 181, vol. 1 (Springer, Berlin, 1972)
63. D.C. Liu, J. Nocedal, On the limited memory BFGS method for large scale optimization. Math. Program. **45**(1–3), 503–528 (1989)
64. D. Loghin, A.J. Wathen, Analysis of preconditioners for saddle-point problems. SIAM J. Sci. Comput. **25**(6), 2029–2049 (2004)

65. D.G. Luenberger, The conjugate residual method for constrained minimization problems. SIAM J. Numer. Anal. **7**(3), 390–398 (1970)

66. C.F. Ma, Q.Q. Zheng, The corrected Uzawa method for solving saddle point problems. Numer. Linear Algebra Appl. **22**(4), 717–730 (2015)

67. K.-A. Mardal, R. Winther, Preconditioning discretizations of systems of partial differential equations. Numer. Linear Algebra Appl. **18**(1), 1–40 (2011)

68. L.D. Marini, An inexpensive method for the evaluation of the solution of the lowest order Raviart-Thomas mixed method. SIAM J. Numer. Anal. **22**(3), 493–496 (1985)

69. H.O. MaY, The conjugate gradient method for unilateral problems. Comput. Struct. **22**(4), 595–598 (1986)

70. S.F. McCormick, *Multigrid Methods*. Frontiers in applied mathematics, vol. 3 (Society for Industrial and Applied Mathematics, Philadelphia, 1987)

71. G.A. Meurant, *Computer Solution of Large Linear Systems*. Studies in Mathematics and Its Applications, vol. 28 (North-Holland, Amsterdam, 1999)

72. J. Nocedal, S. Wright, *Numerical Optimization*. Springer Series in Operations Research and Financial Engineering, 2nd edn. (Springer, New York, 2006)

73. R.W. Ogden, *Non-linear Elastic Deformations*. Ellis Horwood Series in Mathematics and Its Applications (Ellis Horwood, Chichester, 1984)

74. C.C. Paige, M.A. Saunders, Solution of sparse indefinite systems of linear equations. SIAM J. Numer. Anal. **12**(4), 617–629 (1975)

75. J. Pestana, A.J. Wathen, Natural preconditioning and iterative methods for saddle point systems. SIAM Rev. **57**(1), 71–91 (2015)

76. L. Plasman, J. Deteix, D. Yakoubi, A projection scheme for Navier-Stokes with variable viscosity and natural boundary condition. Int. J. Numer. Methods Fluids **92**(12), 1845–1865 (2020)

77. A. Quarteroni, A. Valli, Theory and application of Steklov-Poincaré operators for boundary-value problems, in *Applied and Industrial Mathematics: Venice - 1, 1989*, ed. by R. Spigler, Mathematics and Its Applications (Springer, Dordrecht, 1991)

78. A. Quarteroni, R. Sacco, F. Saleri, *Méthodes Numériques: Algorithmes, Analyse et Applications* (Springer, Milano, 2007)

79. R.T. Rockafellar, The multiplier method of Hestenes and Powell applied to convex programming. J. Optim. Theory Appl. **12**(6), 555–562 (1973)

80. Y. Saad, A flexible inner-outer preconditioned GMRES algorithm. SIAM J. Sci. Comput. **14**(2), 461–469 (1993)

81. Y. Saad, *Iterative Methods for Sparse Linear Systems*, 2nd edn. (Society for Industrial and Applied Mathematics, Philadelphia, 2003)

82. B.V. Shah, R.J. Buehler, O. Kempthorne, Some algorithms for minimizing a function of several variables. J. Soc. Ind. Appl. Math. **12**(1), 74–92 (1964)

83. D.J. Silvester, V. Simoncini, An optimal iterative solver for symmetric indefinite systems stemming from mixed approximation. ACM Trans. Math. Softw. **37**(4), (2011)

84. J.C. Simo, T.J.R. Hughes, *Computational Inelasticity*. Interdisciplinary Applied Mathematics, vol. 7. (Springer, New York, 1998)

85. R. Temam, *Navier–Stokes Equations Theory and Numerical Analysis*. Studies in Mathematics and Its Applications (North-Holland, Amsterdam, 1977)

86. L.N. Trefethen, D. Bau, *Numerical Linear Algebra* (Society for Industrial and Applied Mathematics, Philadelphia, 1997)

87. A. van der Sluis, H.A. van der Vorst, The rate of convergence of conjugate gradients. Numer. Math. **48**(5), 543–560 (1986)

88. H.A. van der Vorst, *Iterative Krylov Methods for large Linear Systems*. Cambridge Monographs on Applied and Computational Mathematics, vol. 13 (Cambridge University Press, New York, 2003)

89. R. Verfürth, A posteriori error estimation and adaptive mesh-refinement techniques. J. Comput. Appl. Math. **50**(1), 67–83 (1994)

90. A.J. Wathen, Realistic eigenvalue bounds for the Galerkin mass matrix. IMA J. Numer. Anal. **7**(4), 449–457 (1987)
91. P. Wriggers, *Computational Contact Mechanics*, 2nd edn. (Springer, Berlin, 2006)

Index

M_S, 11

Active constraints, 24
Active set stategy, 86
Active set strategy, 93
Arrow-Hurwicz-Uzawa, 39, 61
Augmented Lagrangian, 5, 13, 41, 45
    dual problem, 15
    iterated penalty, 17
    discrete, 15

Choice of M_S, 35
    variable coefficients, 57
Coercivity, 46
    on the kernel, 4, 6, 51
Condition number, 12
Conjugate gradient, 22, 23
contact pressure, 88
Contact problems, 75
Contact status, 92
Convergence
    independence of mesh size, 60
Convex constraints, 25

Dirichlet condition, 75
Discrete dual problem, 11
Discrete mixed problem, 7
Discrete norm, 82
Discrete scalar product, 90
Dual problem, 5

Elasticity

choice of elements, 56
Mooney Rivlin model, 52
neo-hookean model, 52
Ellipticity
    global, 8
    on the kernel, 8
Error estimates, 8
Existence, 4

Factorisation, 33
Fractional order derivatives, 77

GCR, 28
GCR solver for the Schur complement, 36
General mixed preconditioner, 34, 38, 56
GMRES, 62

Hierarchical basis, 26

Incompressible elasticity, 48
inequality, 24
Inequality constraints, 24
inf-sup condition, 4

Kuhn-Tucker conditions, 85, 92

Linear elasticity, 49

Matricial form, 9

© The Author(s), under exclusive license to Springer Nature Switzerland AG 2022
J. Deteix et al., *Numerical Methods for Mixed Finite Element Problems*,
Lecture Notes in Mathematics 2318, https://doi.org/10.1007/978-3-031-12616-1

LECTURE NOTES IN MATHEMATICS

Springer

Editors in Chief: J.-M. Morel, B. Teissier;

Editorial Policy

1. Lecture Notes aim to report new developments in all areas of mathematics and their applications – quickly, informally and at a high level. Mathematical texts analysing new developments in modelling and numerical simulation are welcome.

   Manuscripts should be reasonably self-contained and rounded off. Thus they may, and often will, present not only results of the author but also related work by other people. They may be based on specialised lecture courses. Furthermore, the manuscripts should provide sufficient motivation, examples and applications. This clearly distinguishes Lecture Notes from journal articles or technical reports which normally are very concise. Articles intended for a journal but too long to be accepted by most journals, usually do not have this "lecture notes" character. For similar reasons it is unusual for doctoral theses to be accepted for the Lecture Notes series, though habilitation theses may be appropriate.

2. Besides monographs, multi-author manuscripts resulting from SUMMER SCHOOLS or similar INTENSIVE COURSES are welcome, provided their objective was held to present an active mathematical topic to an audience at the beginning or intermediate graduate level (a list of participants should be provided).

   The resulting manuscript should not be just a collection of course notes, but should require advance planning and coordination among the main lecturers. The subject matter should dictate the structure of the book. This structure should be motivated and explained in a scientific introduction, and the notation, references, index and formulation of results should be, if possible, unified by the editors. Each contribution should have an abstract and an introduction referring to the other contributions. In other words, more preparatory work must go into a multi-authored volume than simply assembling a disparate collection of papers, communicated at the event.

3. Manuscripts should be submitted either online at www.editorialmanager.com/lnm to Springer's mathematics editorial in Heidelberg, or electronically to one of the series editors. Authors should be aware that incomplete or insufficiently close-to-final manuscripts almost always result in longer refereeing times and nevertheless unclear referees' recommendations, making further refereeing of a final draft necessary. The strict minimum amount of material that will be considered should include a detailed outline describing the planned contents of each chapter, a bibliography and several sample chapters. Parallel submission of a manuscript to another publisher while under consideration for LNM is not acceptable and can lead to rejection.

4. In general, **monographs** will be sent out to at least 2 external referees for evaluation.

   A final decision to publish can be made only on the basis of the complete manuscript, however a refereeing process leading to a preliminary decision can be based on a pre-final or incomplete manuscript.

   Volume Editors of **multi-author works** are expected to arrange for the refereeing, to the usual scientific standards, of the individual contributions. If the resulting reports can be forwarded to the LNM Editorial Board, this is very helpful. If no reports are forwarded or if other questions remain unclear in respect of homogeneity etc, the series editors may wish to consult external referees for an overall evaluation of the volume.

5. Manuscripts should in general be submitted in English. Final manuscripts should contain at least 100 pages of mathematical text and should always include

   – a table of contents;
   – an informative introduction, with adequate motivation and perhaps some historical remarks: it should be accessible to a reader not intimately familiar with the topic treated;
   – a subject index: as a rule this is genuinely helpful for the reader.
   – For evaluation purposes, manuscripts should be submitted as pdf files.

6. Careful preparation of the manuscripts will help keep production time short besides ensuring satisfactory appearance of the finished book in print and online. After acceptance of the manuscript authors will be asked to prepare the final LaTeX source files (see LaTeX templates online: https://www.springer.com/gb/authors-editors/book-authors-editors/manuscriptpreparation/5636) plus the corresponding pdf- or zipped ps-file. The LaTeX source files are essential for producing the full-text online version of the book, see http://link.springer.com/bookseries/304 for the existing online volumes of LNM). The technical production of a Lecture Notes volume takes approximately 12 weeks. Additional instructions, if necessary, are available on request from lnm@springer.com.

7. Authors receive a total of 30 free copies of their volume and free access to their book on SpringerLink, but no royalties. They are entitled to a discount of 33.3 % on the price of Springer books purchased for their personal use, if ordering directly from Springer.

8. Commitment to publish is made by a *Publishing Agreement*; contributing authors of multiauthor books are requested to sign a *Consent to Publish form*. Springer-Verlag registers the copyright for each volume. Authors are free to reuse material contained in their LNM volumes in later publications: a brief written (or e-mail) request for formal permission is sufficient.

Addresses:
Professor Jean-Michel Morel, CMLA, École Normale Supérieure de Cachan, France
E-mail: moreljeanmichel@gmail.com

Professor Bernard Teissier, Equipe Géométrie et Dynamique,
Institut de Mathématiques de Jussieu – Paris Rive Gauche, Paris, France
E-mail: bernard.teissier@imj-prg.fr

Springer: Ute McCrory, Mathematics, Heidelberg, Germany,
E-mail: lnm@springer.com

Printed in the United States
by Baker & Taylor Publisher Services